BIM in Real Estate Operations

Michael May · Markus Krämer · Maik Schlundt
Editors

BIM in Real Estate Operations

Application, Implementation, Digitalization Trends and Case Studies

Springer

Editors
Michael May
Deutscher Verband für Facility Management (GEFMA)
Bonn, Germany

Markus Krämer
Fachbereich Technik und Leben
Hochschule für Technik und Wirtschaft Berlin
Berlin, Germany

Maik Schlundt
DKB Service GmbH
Berlin, Germany

ISBN 978-3-658-40829-9 ISBN 978-3-658-40830-5 (eBook)
https://doi.org/10.1007/978-3-658-40830-5

This book is a translation of the original German "BIM im Immobilienbetrieb" by May, Michael, published by Springer Fachmedien Wiesbaden GmbH in 2022. The translation was done with the help of an artificial intelligence machine translation tool. A subsequent human revision was done primarily in terms of content, so that the book will read stylistically differently from a conventional translation. Springer Nature works continuously to further the development of tools for the production of books and on the related technologies to support the authors.

This Springer imprint is published by the registered company Springer Fachmedien Wiesbaden GmbH, part of Springer Nature.
The registered company address is: Abraham-Lincoln-Str. 46, 65189 Wiesbaden, Germany

Preface

Building Information Modeling (BIM) has been playing a significant role in the design and construction of buildings for more than two decades.

BIM is a method that is to be considered holistically over the entire lifecycle of a property. A BIM model is the digital image from the planning and construction phase and thus an important data basis for real estate (RE) and facility management (FM) processes in the operational phase. Generally, digitalization is understood to mean the capture and processing of graphical and alphanumeric building data in the BIM model, whereas the holistic BIM concept, including the adaptation of organization and processes, describes the digital transformation as a collaborative method. Since ultimately only the digital transformation leads to a successful use, the term "digitalization" is always used in the sense of digital transformation in this book.

By supporting a cross-disciplinary collaboration and providing the IT tools and platforms necessary for this, BIM can enable faster and more error-free planning resulting in faster construction of buildings.

There is now a wealth of knowledge available on many aspects of BIM. Every specialist book or magazine with a focus on architecture or construction will contain articles and practical reports on the subject. BIM is still rarely considered from the perspective of RE management and FM, however, which was a key motivation for writing this book. The relationships and integration possibilities between BIM tools and IT systems and platforms (Computer Aided Facility Management—CAFM and Integrated Workplace Management Systems—IWMS) used in FM are also rarely exploited. This means that opportunities for a continuous digitalization of real estate and facility management are being missed.

The aim of this book is to help close the gap between BIM-based planning/construction and operation, as well as between the IT tools used for these purposes, and to provide the necessary knowledge.

CAFM/IWMS applications can benefit considerably from BIM. In particular, BIM data, which is often generated during the planning phase of a building, can be used and updated in FM. The consideration of BIM in RE operations and in the broader sense of digitalization of the property industry are essential topics of this book.

The Digitalization Working Group (WG) of the German Facility Management Association (GEFMA) has been dealing with the successful use of information technology (IT) in FM for over 20 years. This includes the development of guidelines, national and international publishing and presentation activities, the identification and dissemination of new digitalization trends, software certification, market and trend studies, writing books and white papers, as well as the organization of their own specialist events such as the Future Lab Digitalization.

The WG therefore became aware of BIM at an early stage. When noticed that BIM was only being adopted and implemented hesitantly in RE management and, in particular, in facility management, and that there were clear knowledge deficits, the WG set up a special working group to investigate BIM from an FM and CAFM perspective. The aim was to disseminate relevant BIM knowledge among facility managers. The result was a German white paper "Building Information Modeling für Facility Manager" (GEFMA 926) in 2017. Since this was very well received by the target group, an update was already available in 2019.

Building on the positive experiences with the white paper, the idea was born to present the topic of BIM in real estate and facility management in more detail. The result is now available with this book. It is based on the structure of the white paper, but has been significantly expanded in terms of content.

A number of fundamental questions are discussed in this book:

- What added value does BIM offer for real estate and facility management?
- What are the most important concepts, terms, standards, data formats, interfaces and technologies that RE and facility managers should be aware of when using BIM in operations?
- How can BIM be used successfully in the operational phase of facilities?
- What benefits can BIM bring to (CA)FM?
- What experiences and suggestions do already implemented BIM/FM projects offer?

An introduction is given to the BIM method, the basic standards and the BIM processes based on them. So real estate and facility managers can assess which challenges and tasks are associated with a BIM project. Subsequently, basic digitalization trends relevant to the real estate industry are presented. These can be used independently or in combination with BIM. It is also clear that the digital transformation in the real estate sector is not to be equated with BIM, but comprises considerably more topics. After introducing the topics of BIM and CAFM, the concept of the digital twin, which is becoming increasingly important for BIM, and the possible IT ecosystems in a BIM project are explained. In addition to the economic efficiency of a BIM project and its evaluation using the Balanced Scorecard method, the procedure for implementing a successful BIM project is considered. The integration of BIM with other areas such as CAFM and Enterprise Resource Planning (ERP) is a frequent requirement in practice and is therefore presented in more detail. Case studies in which BIM is also used in the operational phase form an important focus of the book. The experiences are intended to give suggestions for the implementation of own projects. The book ends with a look at the development of BIM in FM and an insight into corresponding research activities.

The book is aimed in particular at:

- Real estate and facility managers,
- Architects, planners, civil engineers, BIM managers and coordinators who already use the BIM method during planning and construction and want to successfully transfer it to the operational phase,
- BIM/CAFM software providers and implementation partners,
- CAFM/IWMS users who want to connect their systems with BIM tools,
- FM/BIM consultants,
- FM service providers as well as
- Educators and students in real estate-related programs.

Finally, we can only wish that the readers receive valuable suggestions for the successful implementation and utilization of Building Information Modeling in their daily FM activities with the help of this book.

At this point we would like to express our thanks. A special thank you goes to all co-authors and all members of the Working Group Digitalization for their support as well as the many years of constructive and pleasant cooperation. We would also like to thank GEFMA not only for the constant support of the activities of the working group, but in particular this book project.

Furthermore, we would like to thank Michael Huber, Head of Facility Management in the Hochbauamt Graubünden in Switzerland, for his active support in the creation of the BIM2FM case study "sinergia" as well as Björn Erb, Head of Development at Leicom ITEC AG in Winterthur for the description of the ZHELIO research project. The automatic translation by an AI-based tool was thoroughly checked by the editors and adjusted where necessary. Special thanks go to our co-author Dr. Simon Ashworth from ZHAW Zurich for his willingness to do the final review of the book.

We would like to express our sincere thanks to Mrs. Karina Danulat and Mrs. Annette Prenzer from Springer-Verlag for their always pleasant and professional cooperation.

And finally, the editors are obliged to their families for their permanent consideration and support.

Berlin, Germany
Spring 2023

Michael May
Markus Krämer
Maik Schlundt

Simon Ashworth After 23 years of working as an FM and project manager at Serco, Simon Ashworth is now a lecturer at the Institute of Facility Management at the ZHAW in Wädenswil, Switzerland. His research focuses on digital transformation and BIM in FM, real estate and construction. He has a B.Sc. in civil engineering, an M.Sc. in FM and has recently completed his PhD on "The Evolution of Facility Management in the Building Information Modelling Process: An opportunity to Use Critical Success factors for Optimising Built Assets" at Liverpool John Moores University, UK. This was awarded the GEFMA prize in 2021. He is involved in several ISO committees developing BIM and FM standards, as well as in the openBIM professional certification with buildingSMART.

Thomas Bender has been head of products & innovations at pit—cup GmbH since 2019. Here he drives the further development of the pit products towards a sustainable IT ecosystem. The integration of current methods and new technologies such as BIM, IoT or cloud into the pit solutions creates the basis for a digital twin of a building. Previously, he worked for the FM service provider GOLDBECK and then for 14 years at Drees & Sommer in the field of real estate IT consulting.

Through his activities as a member of the GEFMA WG Digitalization and as a board member of the CAFM RING, he actively shapes the development of digitalization topics in the industry. Thomas Bender is the author and co-author of numerous specialist publications such as the CAFM Handbook. As a lecturer for CAFM and BIM2FM, he makes his expert knowledge available to a wide audience.

Nancy Bock After completing a dual degree at the Berlin University of Applied Sciences (2001–2004), Nancy Bock (Dipl. Betriebswirtin/BA) worked for 15 years in various positions in the field of technical building management and facility services. In project controlling, as well as in the management of projects and processes for major customers, her focus was already on the digitalization of order processing and the use of CAFM systems. Since 2020 she has been working at BuildingMinds. There she is responsible for co-innovation with the corporate real estate management of a large chemical company and, as an industry expert, supports the development of a platform for sustainable real estate portfolio management. Nancy Bock is a member of the GEFMA working group on digitalization.

Prof. Dr.-Ing. Asbjörn Gärtner studied civil engineering at the TU Kaiserslautern, with a focus on civil engineering informatics and facility management. As a research assistant, he obtained his doctorate in this field and was instrumental in setting up the first university FM degree program in Germany. After completing his doctorate, he moved to the CAFM vendor KeyLogic as an operational director and then worked as a project manager and consultant at Archibus Solution Center Germany.

As an independent CAFM consultant, he supports companies from all sectors in the implementation of CAFM systems. He is also a lecturer in the field of FM and BIM at various universities. Since January 2021, he has been a professor of facility management at IU International University.

Michael Härtig studied computer science with a focus on technical computer science at the Westsächsische Hochschule Zwickau. He has been working for N+P Informationssysteme GmbH since 1998. He has been involved in the development and implementation of the CAFM system SPARTACUS Facility Management® since working on his diploma thesis. In the past, he has supported CAFM customer projects in various industries.
Today he is the product owner for SPARTACUS and is responsible for the further development and future orientation of the software. Since 2012, he has been a member of the GEFMA working group on digitalization and a co-author of various publications of the working group.

Reiko Hinke leads the BASF building space management at the Ludwigshafen site. His responsibilities include internal rental, the development and implementation of modern office space concepts, the operation and further development of the CAFM software. He studied civil engineering at the Technical University of Darmstadt and was already interested in cost allocations and re-planning in building management systems during his studies. From 1996 he was active in different functions in facility management before he switched to BASF in 2003. He is a member of the GEFMA working group on digitalization.

Univ.-Prof. Dr.-Ing. Dipl.-Inform. Joachim Hohmann studied electrical engineering at the Technical University Darmstadt and computer science at the University Karlsruhe (today KIT). He spent 20 years in senior management positions at the US IT companies Digital Equipment, Hewlett-Packard, and EDS. Since 1996 he has been involved in FM technology with a software vendor and as a management consultant. Since 2002 he teaches as professor for facility management at the Technical University Kaiserslautern. From 2010 to 2013 he was a member of the global board of directors with IFMA (International Facility Management Association), USA. In 2018 he was appointed IFMA Fellow. He is a founding member of the GEFMA working group digitalization and still active as an independent senior industry expert in the field of digitalization of real estate and facility management for public authorities, corporates and consulting firms.

Erik Jaspers has been working in IT for over 40 years. He started his career at Philips Electronics in production automation. For the last 21 years he has been working for the IWMS/CAFM software provider Planon. He has held leading management positions in software development and is currently working on Planon's product strategy and innovation policy.

He has contributed to various publications on IT/FM topics, such as the IFMA books "Work on the Move", "Technology for Facility Managers" and the GEFMA publication "CAFM Handbook". He is co-author of technology articles for magazines such as FMJ and Real Estate Journal and is a speaker at international conferences. He is a member of the IFMA EMEA Board, a member of the IFMA RBI Board and was a member of the GEFMA Working Group on Digitalization until 2022. In 2021 he was awarded the IFMA Fellow status.

Dipl. Inf. (FH), Dipl. Facility Manager (GEFMA) Thomas Kalweit studied at the FHTW Berlin from 1998 to 2002. After working as a self-employed person, he first worked as a technical assistant and then as an IT/CAFM project manager and managing director in the FM industry. From 2015 to 2019 he was responsible for the implementation of CAFM projects as divisional manager of Ambrosia FM Consulting & Services GmbH. Since 2019 Thomas Kalweit has been working as head of "Development and Innovation" for net-haus GmbH. His tasks include, among other things, the further development of existing and the development of innovative software products for the real estate industry.

Thomas Kalweit has been a member of the GEFMA Working Group "Digitalization" since 2009 and has intensively been involved in the creation of various guidelines, the white paper "Cloud Computing in FM" and the publication "CAFM Handbook".

Dr.-Ing. Stefan Koch studied mechanical engineering at the TU Berlin. In 1993 he obtained his doctorate there on IT-supported automation processes. From 1986 to 1994 he was a scientific employee at the Fraunhofer Institute for Production Systems and Construction Technology (IPK), subsequently consultant at A. T. Kearney. Since 1995 Stefan Koch has been managing director of Axentris Informationssysteme GmbH. Axentris is the manufacturer of the Servalino software, which supports both the commercial and the technical processes around construction, operation and use of real estate.

Since 2001 he has been a member of the GEFMA working group on digitalization. In the Society for the Promotion of Applied Informatics (GFaI) he has been a member of the board of directors since 2002.

Dr. Bernd Limberger studied construction operations and construction economics at the University of Karlsruhe and obtained his doctorate from Prof. C. J. Diederichs in Wuppertal. Since 2001 he has been working as a consultant in the field of IT support for real estate management processes. In 2007 he moved to SAP Deutschland in the consulting team for real estate. There he had various roles as consultant, project manager, trainer and business development manager. Since 2016 he has been head of consulting for the real estate management sector. Within the global consulting organization of SAP, he also coordinates the transfer of knowledge from the German consulting team for real estate management to the worldwide sister teams. Bernd Limberger has been a member of the GEFMA working group on digitalization since 2020.

Dipl.-Math. Michael Marchionini has been working in facility management since 1994. The focus of his work was on the conceptual preparation of CAFM projects and on assisting companies during CAFM implementation. Since 2008, his focus has been on the strategic planning of office space portfolios using IT-supported methods. As managing partner of the company he founded ReCoTech GmbH, where he gained experience in the planning of over 2 million m^2 of office space in key projects, also in the role of project manager. He documents these experiences as part of his work on the GEFMA working group on digitalization, inter alia, as author or co-author of the GEFMA guidelines 400 ff on CAFM and 130 on space management. In addition, he is co-author of the books "Flächenmanagement in der Immobilienwirtschaft" and "CAFM-Handbuch" published by Springer-Verlag.

Matthias Mosig is a managing director and Head of Digital Transition at TÜV SÜD Advimo GmbH. As a civil engineer, he was initially active in construction project management and switched to real estate and facility management consulting about 20 years ago. As a co-founder of cgmunich GmbH, in addition to his management activities, he was mainly active in process and IT consulting. After the sale of the company to TÜV SÜD, he promoted the development of BIM (Building Information Modeling) consulting and BIM management in his area of real estate consulting until he took over the position of Head of Digital Transition. In this role, he is an active member of various working groups of leading industry associations. He has been heading the GEFMA working group on digitalization since 2022.

Marko Opić During his studies of supply engineering, Dipl.-Ing. (FH) Marko Opić already accompanied projects in the fields of integrated supply systems and facility management at Ebert-Ingenieure in Nuremberg, where he started his professional career as an FM consultant and QM officer in 2001. After stations at VALTEQ and CBRE, he changed to Alpha IC in 2017 as a senior project manager. His consulting focus is on planning-accompanying FM consulting, operational concepts, service provider selection and control, usage cost determination, as well as IT concepts in FM. Since the company switched to an agile management system, he has been a partner in the management and is responsible for an interdisciplinary and location-spanning consultant team. Marko Opić has been a member of the GEFMA working group "Digitalization" since 2002.

Nino Turianskyj completed his studies in information technology at TU Chemnitz-Zwickau from 1988 to 1994 with the degree of Diplom-Ingenieur. He then began working in the engineering company Keßler or the later Keßler Real Estate Solutions GmbH. Since 1997 he has been working in the field of FM/CAFM and has accompanied numerous implementation projects and was responsible for the development of the CAFM software FAMOS as head of development. From 2008 to 2020 he was a member of the GEFMA working group "Digitalization". In his spare time he worked on several books on programming as well as on the CAFM handbook. Since 2021 he has been working independently in the field of LED lighting. He is also involved in the Kanimambo e.V. association in an educational project for children in Mozambique.

1 The Built Environment, BIM and the FM Perspective

Simon Ashworth and Michael May

Buildings and their facilities form one of the most important foundations of modern society. They touch every aspect of our lives and provide commercial offices, stores, hospitals, schools, transportation networks, etc. for public and private organizations. Wilson (2018) notes their importance in that they make up a considerable part of the world's wealth and a large proportion of the worlds workforce is employed in this sector. Buildings are essential for sustainable development, as highlighted in the "2030 Agenda for Sustainable Development" (NN 2015a), which outlines 17 sustainability goals (Sustainable Development Goals, SDGs), as shown in Fig. 1.1 (NN 2015a). These are divided into 169 sub-goals, of which almost three quarters relate directly to infrastructure (Adshead et al. 2019). These goals aim to address urgent and critical challenges that our planet is currently facing.

The challenges are driven by the growth of the world population, which is expected to reach 9.7 billion by 2050 (NN 2019b). Human activities have placed a heavy burden on the limited natural resources of our planet and the United Nations predicts that if this is not controlled, it will lead to irreversible damage to our planet (NN 2019c). SDG 11 specifically deals with the importance of "sustainable cities and communities". It states that since 2007 more than half of the world's population has lived in cities and that this proportion will increase to 60% by 2030 (NN 2015b). The built environment plays a key role as it is responsible for 60–80% of energy consumption, 75% of carbon emissions and about 60% of urban waste. Therefore, improvements in this area can have a signifi-

S. Ashworth (✉)
Zürcher Hochschule für Angewandte Wissenschaften (ZHAW), Wädenswil, Switzerland
e-mail: ashw@zhaw.ch

M. May
Deutscher Verband für Facility Management (GEFMA), Bonn, Germany
e-mail: michael.may@gefma.de

M. May et al. (eds.), *BIM in Real Estate Operations*,
https://doi.org/10.1007/978-3-658-40830-5_1

Fig. 1.1 UN Sustainable Development Goals

cant impact on reducing waste and environmental pollution, improving energy efficiency, and improving the living conditions and quality of life of people.

Construction and facility management (FM) are also of key importance for economic growth. ResearchAndMarkets (NN 2021ac) estimated growth of the global market from $11,491.42 billion in 2020 to $12,526.4 billion in 2021. These industry sectors employ about 7% of the world's working population and accounts for about 13% of global GDP. According to forecasts, the FM industry will reach $1,759 billion globally by 2028 (NN 2021ac). The significant impact on the environment and the financial influence on the world economy oblige the construction and FM industry to proactively drive change in order to support humanity in achieving the SDG goals.

1.1 Digital Transformation of Construction, Real Estate and Facility Management

The construction industry has recorded some of the world's lowest annual productivity growth rates over the last 20 years, an important metric for measuring the performance of the sector. Barbosa et al. (2017) showed an increase of 1% for the construction industry, while other sectors such as manufacturing achieved 3.6%. Baller et al. (2016) found that digitalization, along with other industrial, scientific and technical advances, is key to transformation in the sector. This is where the greatest hope lies for addressing the United Nations' SDG challenges. It is driven by new conditions for knowledge and information exchange and technological development. Technological advances include, for example, a 10,000-fold increase in computing power since 2000 and a 3000-fold reduc-

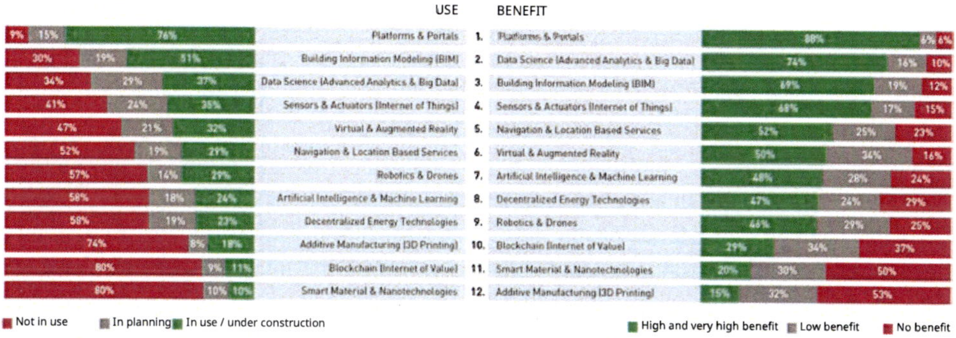

Fig. 1.2 Maturity of digital technologies in the construction and real estate industry

tion in storage costs (Menon 2018). The technological revolution often referred to as Industry 4.0 has led to a global digital transformation affecting all sectors of industry. The number of digitally connected devices has exploded and is expected to reach 75 billion by 2025 (NN 2021ae).

Studies by Baldegger et al. (2021) on the Swiss and German real estate markets show that most companies are already investing proactively in digitalization, as shown in Fig. 1.2. The pom+ study shows how secure digital technologies are and how companies are using or planning to use them. Building Information Modeling (BIM) is clearly highlighted as one of the most important trends with significant potential seen in it.

1.2 Linking BIM with Sensors

As explained in Sect. 4.1, the next logical step in digitalization is the creation of digital twins. These go far beyond a conventional CAD model and link static data of a BIM model with all its facets with dynamic data supplied by sensors and other devices from buildings and properties (see also Sect. 2.1). This makes it possible to provide dynamic real-time data about buildings and facilities. This enables time-based analysis and dynamic adaptation of performance or conditions based on that data (Wright and Davidson 2020). Digital twins extend BIM models; but they can also be based on simpler 3D representations of buildings created with other digital capture technologies such as laser scanning or photogrammetry (see Sect. 5.2). They enable predictive modeling for proactive optimization of real estate. Perhaps even more important, they give people the ability to interact directly with buildings and help measure and improve the well-being of building occupants (Fruchter 2021).

1.3 The Importance of BIM During the Lifespan of Buildings

Sawhney (2015) describes BIM as fundamental to the digital transformation of industry. One key reason for this is that BIM can and should be used across all lifecycle phases, from initial procurement through planning, design, construction and into the much longer operational phase and final disposal or repurposing. Fig. 1.3 (Ashworth 2021) illustrates why it is important to focus on the operational expenses (OPEX) phase, which comes on top of the initial capital expenses (CAPEX) during the much longer use phase, and then represents the majority of the total costs. Focusing on the long-term OPEX costs provides a far greater chance of achieving good value for money and improving sustainability aspects, for example through energy savings and waste reduction.

However, studies by Baldegger et al. (2021) in Switzerland and Germany show that stakeholders still assess the value and benefits of BIM differently. Contractors and suppliers already see 100% benefits from BIM for their tasks, while other stakeholders report lower percentages (see Fig. 1.4). If BIM is to be universally implemented and used, these numbers need to increase.

The lower numbers, in particular for investors, FM, portfolio and building owners, may come as a surprise at first, especially since Eadie et al. (2013) point out that owners and facility managers benefit most from the introduction of BIM. The National Building Specification (NBS) in the UK (NN 2020d) notes that the client has a significant (and probably the most important) role to play within the information management ecosystem. However, it is also noted that the most common obstacle (64%) to the introduction of BIM is the lack of requirements from customers. Experience shows that those who use BIM as part of their daily work adopt BIM more quickly and realize its potential. Building owners and facility managers, who generally use the project result, in particular the information, need to become more familiar with the potential benefits (see Sect. 1.4, 3.3 and 6.3) and understand how BIM can contribute to the strategy of their organization in order to better manage their real estate portfolio and the associated services.

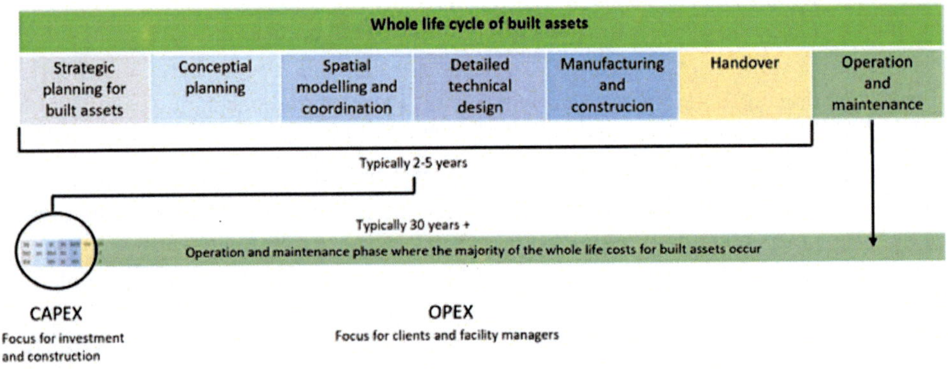

Fig. 1.3 The importance of the operating phase within the lifecycle of built assets

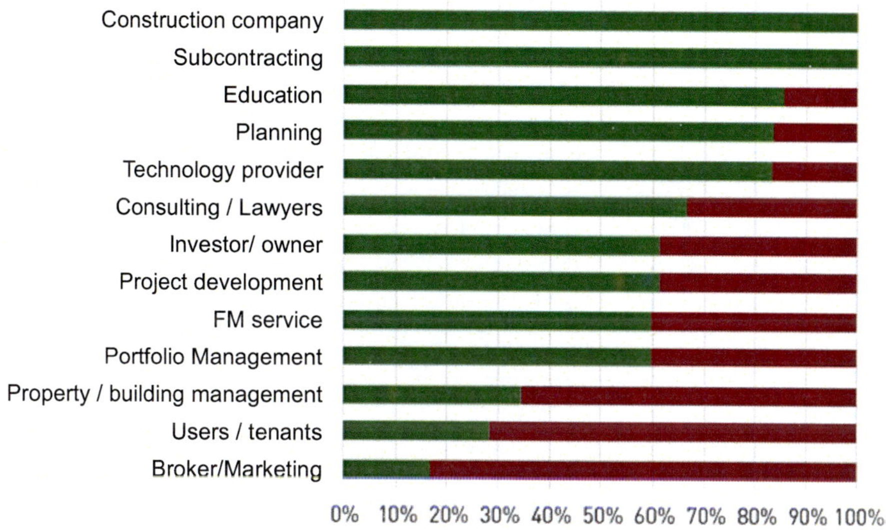

Fig. 1.4 Importance of BIM for different stakeholders

1.4 Benefits of BIM in Real Estate Operations

The benefits of BIM for the planning, design and construction of buildings are well documented. However, Sacks et al. (2018) note that most owners are not fully aware of the benefits that BIM can offer during the long operational phase. These are extensive but often intangible and difficult to quantify because buildings are only built once and it is difficult to compare how a project would fare with or without BIM. Some examples of important benefits with direct impact on FM from the literature are summarized in Table 1.1 by Ashworth (2021). Further benefits are discussed in Sects. 3.3 and 6.3.

In a study by Ashworth et al. (2019) 373 occurences of benefits of BIM for FM found in the literature were examined. These were divided into nine main groups, which were ordered by frequency of occurrence (cf. Table 1.2). "Time saving" was the most frequently mentioned benefit category here.

Of central interest to owners and investors is the potential Return on Investment (ROI) when investing in the implementation and use of BIM (see also Chap. 6). As already mentioned, accurate predictions are difficult because each project is unique. However, some examples are the calculations of Teicholz (2013). These show that the possible total savings per year for a typical project are 3.93% and amortize in 1.56 years. The report "PwC BIM Level 2 Benefits Measurement" (NN 2018e) describes two specific examples: the project "39 Victoria Street office refurbishment", for which a total of 3.0% of the costs were saved compared to the costs "without BIM", and the project "Foss Barrier Upgrade" with a total of 1.5% savings.

Table 1.1 Examples of basic benefits of BIM in FM

Benefits	Explanation
Time and efficiency	BIM can reduce project execution time, enable faster and more effective processes, information can be exchanged and reused more easily and have added value.
Performance and analysis	BIM can enable analyses to improve building performance.
Cost savings	BIM can reduce downtime and associated costs by enabling faster response times to emergency work orders.
Energy efficiency	BIM can help to reduce annual energy consumption and minimize environmental risks.
Increase business value	BIM can help reduce the probability of asset downtime due to more accurate understanding of asset condition and avoiding component failures due to timely maintenance.
Data accuracy and quality	BIM can enable better management and organization of information, reduce inaccurate and incomplete information, improve lifecycle planning, as well as durability and sustainability.
Interoperability	The exchange and transfer of BIM data reduces the need for major repairs and alterations, and increases the efficiency of work orders and decision-making processes by access to real-time and previously stored graphical and alphanumeric data.

Table 1.2 Benefits of BIM for FM

Ranking of frequency	Type of benefit category	Percentage
1	Time savings	21.98%
2	Productivity	18.23%
3	Cost savings	16.62%
4	Business values	14.21%
5	Data accuracy and quality	11.26%
6	Communication and collaboration	7.77%
7	Energy performance	4.02%
8	Improvement of safety and risk management	3.75%
9	Interoperability	2.14%
	Total	100.00%

From the FM perspective, BIM projects are only successful if the project team successfully and completely hands over all essential inventory documents, 3D models, and alphanumeric data that are crucial for supporting the customer's strategic and everyday FM processes (NN 2017d). Kensek (2015) stated that a fundamental goal of BIM is to enable the easy exchange of data between and within CAFM systems (Chap. 5). If this is done professionally, then accurate information can be provided that will serve to improve the operation and maintenance of the property and its systems throughout their life cycle (Ashworth et al. 2020).

1.5 Successful Projects Through BIM Strategy and Knowledge Transfer

In order to take full advantage of digitalization and, in particular, of BIM, companies need to develop a sustainable BIM strategy. For this, a clear understanding of what can be achieved with BIM is important. This way, expectations of BIM can be reasonably controlled. It is important to formulate transparent goals with realistic specifications that can also be achieved. Teicholz (2013) states that these goals must be aligned with the company's overall strategy and support the organization's existing real estate and FM strategy in order to maximize benefits. Many companies are currently creating a BIM Roadmap to help employees better understand their own approach in a BIM project as well as the challenges and changes associated with it. Fig. 1.5 (NN 2021ad) shows an example from SBB in Switzerland.

Such strategic development requires appropriate and suitable investments in equipment and BIM training to ensure that teams are able to competently participate in BIM projects and achieve successful results. This was found by Amuda-Yusuf (2018) in the investigation of critical success factors (CSF) in BIM projects. "Training and further

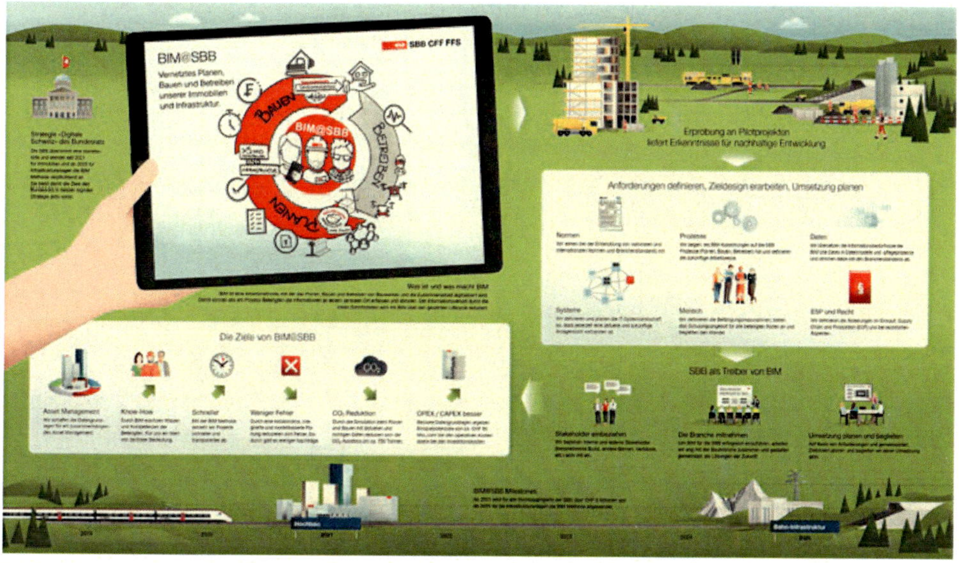

Fig. 1.5 BIM Roadmap of SBB in Switzerland

education" was classified as the third most important factor in the BIM process. Another important aspect that must be examined before starting a BIM project is the question of which data environment (Common Data Environment – CDE) (see also Sect. 4.3) should be used for the management of the entire information and data. Also, binding contract agreements should be made which aim to avoid potential conflicts and to ensure clear ownership of and access to data and information (Saxon et al. 2018).

Some successful BIM practice projects in FM are presented in Chap. 9.

1.6 The Role of FM in BIM Projects

Thomas (2017) points to the key role of facility managers in successful BIM projects. They understand what types of information and data are needed in the operational phase and how the properties and facilities can directly support the overarching corporate strategy. They represent the interests of the owner, the customer (employer) and the end user to ensure that properties and facilities can be operated, maintained and managed effectively. Fig. 1.6 shows an "FM-BIM Strategy Concept Model" developed by Ashworth 2016 and updated in 2020, in which the facility manager acts as the "representative of

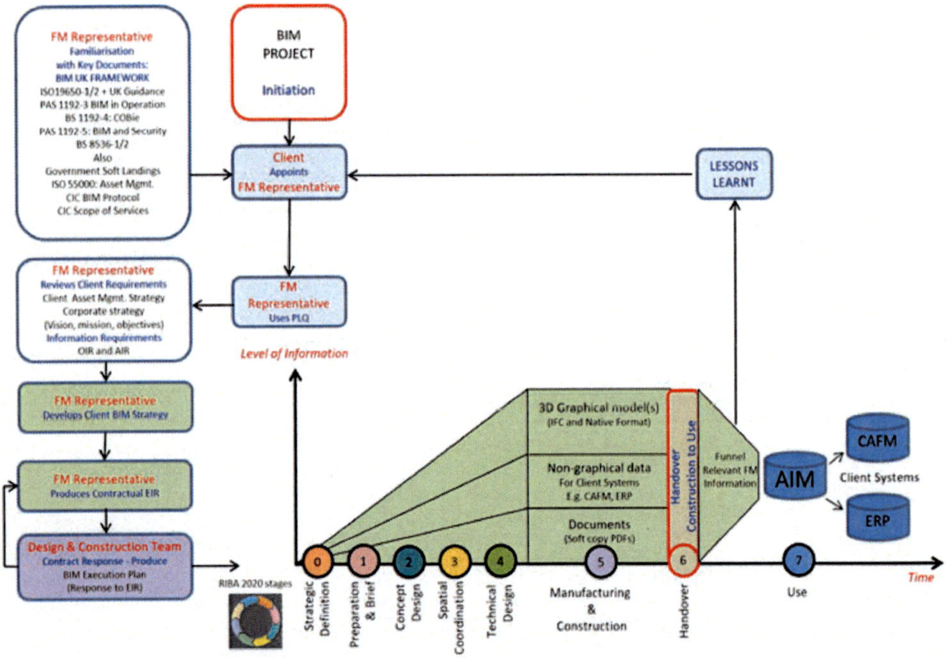

Fig. 1.6 FM-BIM Strategy Concept Model

the customer" (Ashworth et al. 2016). This role is best performed by the internal facility manager who knows the organization very well, but it could also be taken over by a BIM-competent FM consultant. Their contribution is crucial to help the BIM project team understand the information that is needed to support the customer's business processes. Whoever takes on this task must be familiar with the ISO 19650 standards, which provide important guidance for managing information throughout the entire lifecycle. They must also speak the language of FM and construction industries. The model in Fig. 1.6 is based on a UK context, but can be transferred to other countries without any problems. It refers to the "UK BIM Framework" (NN 2021af), which provides a central repository for BIM guidance and support in the UK industry.

The main aim is to ensure that the BIM process provides all relevant information and data needed to support strategically important business processes, e.g. maintenance, asset exchange, health and safety. The information must be collected in a coordinated way by the project team (delivery team according to ISO EN DIN 19650) during project development and be able to be transferred easily or linked with the relevant management systems of the organization, e.g. CAFM, BMS and ERP. In order to ensure that the project team can successfully provide this information, ISO 19650 (NN 2018c) requires that the client defines the information requirements for the project partners (contractors) in several key documents (Sect. 3.4):

- Organizational Information Requirements (OIR):
  These describe the customer's or owner's high-level and strategic information needs (i.e. what is needed for reporting and executing core business activities).
- Asset Information Requirements (AIR):
  These describe the specific information needs required to enable FM and operations teams to manage and optimize the operational efficiency of buildings and assets.
- Employer Information Requirements (EIR):
  These build on the OIR/AIR and should be part of the formal contract with the contractor to define administrative, commercial and technical specifications that instruct the contractor on the information needs and procedures for a specific project.

These documents form the basis for the entire BIM process and must clearly define what information is needed and why. This requires the active involvement of key operational staff with a detailed understanding of the business requirements and processes of the company. Their contribution provides the crucial insight into the "pains and gains" of an organization and what can be improved or optimized. This also includes the determination of IT systems and the BIM information and data that is to be used. This gives the project team a clear understanding of why certain information is needed and in what format.

1.7 Information Requirements—A Critical Success Factor for BIM Projects

The long-term goal of BIM is to ensure that customers who invest a lot of money receive buildings and facilities that not only meet their daily needs, but can also be operated and maintained simply and efficiently over their entire life cycle. Ashworth (2021) points out how important it is to involve FM early on so that the information requirements can be clearly defined at an early stage of the process. Otherwise, the information transfer often takes place inefficiently and will probably be error-prone, which usually leads to poorly structured information and data that are difficult to transfer to CAFM systems or cannot be used productively (May 2018a). This results in additional steps that are time-consuming and expensive.

This problem is not new and was previously described by Gallaher et al. (2004). They pointed to the high costs of a faulty transfer. Even for the automated transfer of information into CAFM tools, $613 million were incurred. Owners and operators have the highest interoperability costs of all stakeholders: more than $10.6 billion or about 68% of the total costs incurred for the supply chain of investment goods are attributable to inadequate interoperability. 85% of the interoperability costs of owners and operators fall in the operating and maintenance phase. The quantified costs were estimated at around $9 billion in 2002.

Early planning also allows for the gradual capture of information as the project progresses. This is especially important for the project team, which is often under pressure to complete everything on time in the final phase of a project. They will be less focused on information gathering at this stage and more on completing the project on time. The definition of clear requirements enables the main contractor to instruct their suppliers clearly to collect and transfer the valuable information and data. If this does not happen, the customers are likely to be disappointed at the end when they do not find the essential information they need in their operational IT systems (Ashworth et al. 2018).

The Fig. 1.7 (cf. Ashworth et al. 2019) underlines the need for an early involvement of FM. The upper arrows illustrate that the handover from "construction" to "operations" should not be considered as a single point in time, but as a stepwise planned process. The left side (in green) stands for the capture of construction information by the handover team: 3D graphics, alphanumeric data and documents. The decreasing green lines during handover illustrate that the knowledge of the handover team decreases and gradually disappears with the handover and the transition to the next project. The right side (in red) represents the team for the most important FM activities. The stepwise, but early involvement of FM is crucial for the definition of information requirements and the support of the project team. This should be expanded during the development of the project and at the time of handover, the FM team should already be well prepared to use the information in daily operation. This includes, for example, processes, services, costs and products. The blue boxes represent the most important BIM process steps (the infor-

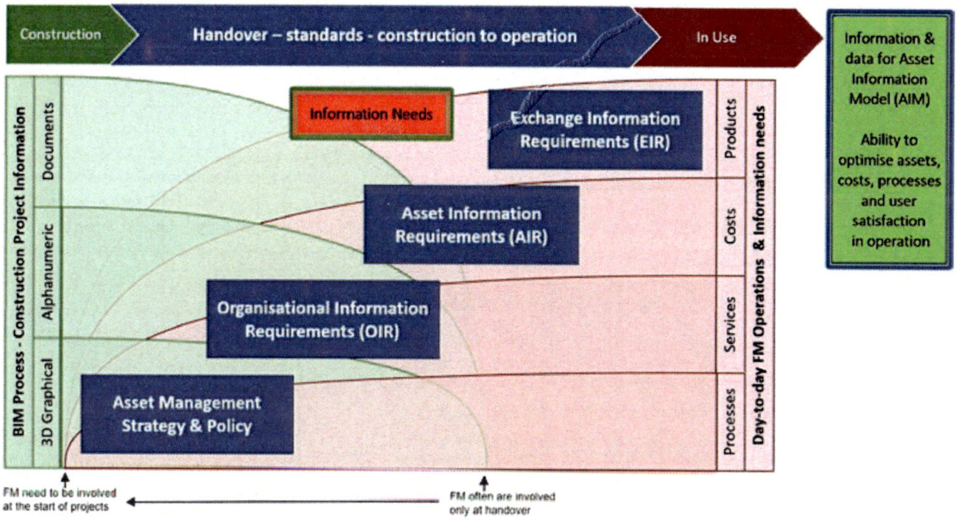

Fig. 1.7 Need for early involvement of FM in the BIM process

mation requirements documents) that must be in place. By aligning with the company's asset/real estate management strategy, good OIR, AIR and EIR can contribute to a successful handover. The final goal, highlighted in the right green box, is to provide the right information and data. This is referred to as the Asset Information Model (AIM) which supports the optimization of real estate, costs, processes and user satisfaction in operation.

1.8 CDE Usage for Information Delivery to the FM Team

In his essay "Content is King" from 1996, Bill Gates stated that "the internet is already revolutionizing the exchange of specialized scientific information". He suggested that Internet users expect to be "rewarded with deep and extremely up-to-date information that they can explore at will" (Evans 2017). Similarly, property and facility managers need current and easily accessible information if they are to effectively manage their properties and facilities, and the services associated with them, throughout their lifetimes. Future technological solutions will include significantly more automated functions that combine knowledge management with BIM to make useful information and data easier to find and evaluate. However, Hu et al. (2021) point out that the reliable use of these capabilities is still in an early stage of development (see also Besenyöi and Krämer 2021). Facility managers need all information about the as-is state for transfer into or use in the AIM of their customer, which may comprise several enterprise-wide information management systems.

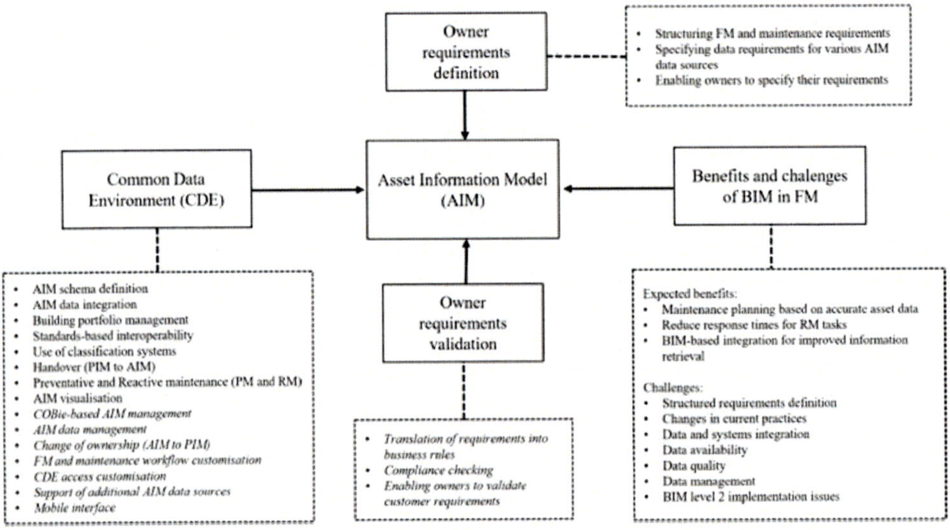

Fig. 1.8 Action framework for an AIM supported by a CDE

Fig. 1.8 (see Patacas et al. 2020) illustrates a possible action framework for setting up an AIM using a Common Data Environment (CDE), based on an open standard. The lists in the dashed boxes highlight some of the key and critical aspects that relate to the definition of client requirements and the CDE. This includes, for example, the use of classification systems and the standard COBie (Construction Operations Building Information Exchange), (see NN 2021e and Sect. 5.3). It also highlights some of the expected benefits and challenges of BIM in FM and, in particular, the validation of owner requirements.

1.9 Validating the Owner Requirements

A central challenge in handover is verifying that the required information and data have been delivered. An interesting question is whether main contractors, who lead the projects and are familiar with BIM, are able to obtain the required information from subcontractors and, in particular, from smaller suppliers. These are often inexperienced in using BIM software and sometimes have difficulty delivering the required information correctly and on time. In general, main contractors receive the information in a variety of formats and bundle this for the client. However, this process is prone to data loss and often results in the client receiving poorly organized and structured information that is difficult to process and use.

Fig. 1.9 shows an innovative solution to overcoming this problem. This allows clients to define their information needs early on. Smaller suppliers can then be given a link that allows them to view the BIM models (without expensive software) and upload the data and documents required for operation. The LIBAL software was developed with the aim

Fig. 1.9 Example of LIBAL tool to check the status of information delivery according to ISO 19650-2

of representing the BIM process described in ISO 19650-2 (NN 2018d). It enables the BIM or information managers of the project to track the status of information delivery for any organization that needs to provide information for the project at any time.

1.10 Strategies for the Digitalization of an Entire Real Estate Portfolio

It is repeatedly pointed out that BIM offers, or can offer a Single Source of Truth. This source should include a wide range of useful information, e.g. components, material types, ground properties, building functions, floor plans, facilities, equipment lists, connections and relations between equipment, product data sheets, warranties and preventive maintenance schedules (Florez and Afsari 2018). However, even if this information may already exist in individual BIM projects, the real estate portfolio of most organizations consists of a wide range of buildings and facilities, for which digital building models may only exist in individual cases at best.

In case of most of the real estates the existing buildings are of different ages with very different information levels. New BIM projects usually only make up a small percentage of the total portfolio. For all other buildings, a time-consuming and costly subsequent data capture is usually required. Depending on the respective building types and business uses, however, simplified digital data capture methods such as laser scanning and photogrammetry (see also Sect. 5.2) or even combinations of different approaches can be used.

Supporting real estate portfolios with BIM requires even more than for individual projects a standardized, uniform BIM strategy (Aengenvoort and Krämer 2018), but then offers the possibility to query information for several objects of the portfolio if the structure of the building models is the same. In addition, real estate portfolios are suitable for the use of template models, at least if part of the buildings have structural similarities. This then also reduces the effort for the BIM capture of existing buildings.

1.11 BIM and Interoperability

FM organizations use a wide range of IT systems to manage buildings, including CAFM/IWMS, CMMS, BMS, BAS, and ERP, which are used throughout the life cycle of buildings. Teicholz et al. (2013) found that a key success factor for BIM projects is exchanging data between BIM and these systems. This requires careful consideration of how the generated BIM data and information should be used and transferred to the various partner systems, or what interfaces and technologies should be used to connect these systems. Without careful planning, problems with interoperability (see also Sects. 3.4, 4.3 and 5.3) of such systems can occur, which can lead to additional costs. The exchange and transfer of data between BIM models and other software is usually done via Industry Foundation Classes (IFC). The IFC format (NN 2018a) provides a snapshot of a model and data that can be transferred to other software systems. However, changes to the model can often not be made in IFC, but are usually made again in the original BIM authoring system software, which is a certain limitation of the IFC format. COBie is also used and allows for the unidirectional export of data from a model with defined criteria, which can be used, for example, in CAFM systems. However, only what is already included in the BIM model can be exported. This underscores the importance of careful planning to decide which aspects (in particular from the FM perspective) should be added to a model. It may make sense not to consider certain aspects in order to keep costs under control or where the information or data is held in another system. It must also be thought about how the BIM models and information should be maintained for future use. This is essential in order to be able to derive lasting benefits from BIM.

1.12 Summary

In this introductory chapter, the importance of the construction, real estate and FM industry is explained in relation to the world economy, sustainability and achievement of the United Nations' Sustainable Development Goals. These industries are all undergoing a digital revolution, with BIM being of central importance for the digitalization of the built environment and the processes associated with it. Another important topic is the creation and further development of digital twins.

In order to overcome the challenges and exploit the potential benefits of BIM in operation, clients and facility managers as well as other stakeholders must be involved in the BIM process at an early stage. They must develop a clear BIM strategy ensuring that information requirements are aligned with the general corporate strategy. These requirements must be developed on the basis of the FM teams' know-how. In addition, there must be clear plans to transfer BIM data seamlessly into operational IT systems or to exchange them reliably with these systems using suitable interfaces. This helps to ensure that the data are available throughout the lifecycle of buildings for a variety of uses.

The topics briefly mentioned here will be explained in more detail in the following chapters and supplemented with suitable examples.

References

Adshead D, Thacker S, Fuldauer LI, Hall J (2019) Delivering on the Sustainable Development Goals through long-term infrastructure planning. Global Environmental Change 59(2019) 1–14

Aengenvoort K, Krämer M (2018) BIM in the Operation of Buildings. In: Borrmann A, König M, Koch C, Beetz J (Eds.) Building Information Modeling – Technology Foundations and Industry Practice. Springer Nature, 2018, 477–491

Amuda-Yusuf G (2018) Critical Success Factors for Building Information Modelling Implementation. Construction Economics and Building 18(2018)3, 55–74

Ashworth (2021) The Evolution of Facility Management (FM) in the Building Information Modelling Process: An opportunity to Use Critical Success factors (CSF) for Optimising Built Assets. Doctoral Thesis, Liverpool John Moores University, UK

Ashworth S, Carey D, Clarke J, Lawrence D, Owen S, Packham M, Tomkins S, Hamer A (2020) BIM Data for FM Systems: The facilities management (FM) guide to transferring data from BIM into CAFM and other FM management systems. https://www.iwfm.org.uk/resource/bim-data-for-fm-systems.html?parentId=4D64E6F8-D893-4FF1-BABA5DF2244A7063 (retrieved: 14.10.2021)

Ashworth S, Druhmann C, Streeter T (2019) The benefits of building information modelling (BIM) to facility management (FM) over built assets whole lifecycle. 18th EuroFM Research Symposium, Dublin, Ireland

Ashworth S, Tucker M, Druhmann C (2016) The role of FM in preparing a BIM strategy and Employer's Information Requirements (EIR) to align with a client's asset management strategy. European Facility Management Conference, Milan

Ashworth S, Tucker M, Druhmann C (2018) Critical success factors for facility management employer's information requirements (EIR) for BIM. Facilities 37(2018)1/2, 103–118

Baldegger J, Gehrer I, Ruppel R, Wolters K, Glättli T, Jost A (2021) pom+ Digitalisierung der Bau- und Immobilienwirtschaft: Digital Real Estate Umfrage 2021. https://www.digitalrealestate.ch/products/digitalisierungsindex-2021 (retrieved: 14.10.2021)

Baller S, Dutta S, Lanvin B (2016) The Global Information Technology Report 2016: Innovating in the Digital Economy. http://www3.weforum.org/docs/GITR2016/WEF_GITR_Full_Report.pdf (retrieved: 14.10.2021)

Barbosa F, Woetzel J, Mischke J, Ribeirinho MJ, Sridhar M, Parsons M, Bertram N, Brown, S. (2017) Reinventing construction: a route to higher productivity: Executive Summary. https://pzpb.com.pl/wp-content/uploads/2017/04/MGI-Reinventing-Construction-Full-report.pdf (retrieved: 14.10.2021)

Besenyöi Z, Krämer M (2021). Towards the Establishment of a BIM-supported FM Knowledge Management System for Energy Efficient Building Operations. Proc. of the 38th International Conference of CIB W78, Luxembourg, 13–15 October, 194–203. http://itc.scix.net/paper/w78-2021-paper-020

Eadie R, Browne M, Odeyinka H, McKeown C, McNiff M (2013) BIM implementation throughout the UK construction project lifecycle: An analysis. Automation in Construction 36(December 2013), 145–151

Evans H (2017) "Content is King" – Essay by Bill Gates 1996. https://medium.com/@HeathEvans/content-is-king-essay-by-bill-gates-1996-df74552f80d9 (retrieved: 14.10.2021)

Florez L, Afsari K (2018) Integrating Facility Management Information into Building Information Modelling using COBie: Current Status and Future Directions. Proc. 35th Int. Symp. on Automation and Robotics in Construction (ISARC 2018), Berlin, 8 p

Fruchter R (2021) When 21st Century Technologies Meet the Oldest Engineering Discipline, Presentation at 38th International Conference of CIB W78, Luxembourg

Gallaher MP, O'Connor AC, Dettbarn JL, Gilday LT (2004) Cost Analysis of Inadequate Interoperability in the U.S. Capital Facilities Industry. https://www.nist.gov/node/583921 (retrieved: 14.10.2021)

Hu Z-Z, Leng S, Lin J-R, Li S-W, Xiao Y-Q (2021) Knowledge Extraction and Discovery Based on BIM: A Critical Review and Future Directions. Archives of Computational Methods in Engineering (April 2021) 22 p

Kensek K (2015) BIM Guidelines Inform Facilities Management Databases: A Case Study over Time. Buildings, 5(August 2015)3, 899–916

May M (Ed.) (2018a) CAFM-Handbuch – Digitalisierung im Facility Management erfolgreich einsetzen. 4. edn., Springer Vieweg, Wiesbaden, 2018, 713 p

Menon P (2018) An Executive Primer to Deep Learning. https://medium.com/@rpradeepmenon/an-executive-primer-to-deep-learning-80c1ece69b34 (retrieved: 14.10.2021)

NN (2015a) Transforming our World: The 2030 Agenda for Sustainable Development. https://sustainabledevelopment.un.org/content/documents/21252030%20Agenda%20for%20Sustainable%20Development%20web.pdf (retrieved: 14.10.2021)

NN (2015b) Sustainable cities: why they matter. https://www.un.org/sustainabledevelopment/wp-content/uploads/2016/08/11.pdf

NN (2017d) Asset Information Requirements Guide: Information required for the operation and maintenance of an asset, 53 p. http://www.abab.net.au (retrieved: 14.10.2021)

NN (2018a) ISO 16739-1: Industry Foundation Classes (IFC) for data sharing in the construction and facility management industries Part 1: Data schema. International Organization for Standardization, 2018-11

NN (2018c) ISO 19650-1:2018: Organization and digitization of information about buildings and civil engineering works, including building information modelling (BIM) – Information management using building information modelling: Part 1: Concepts and principles. https://www.iso.org/standard/68078.html (retrieved: 14.10.2021)

NN (2018d) ISO 19650-2:2018: Organization and digitization of information about buildings and civil engineering works, including building information modelling (BIM) – Information management using building information modelling: Part 2: Delivery phase of the assets. https://www.iso.org/standard/68080.html (retrieved: 14.10.2021)

NN (2018e) BIM Level 2 Benefits Measurement, Application of PwC's BIM Level 2 Benefits Measurement Methodology. https://www.cdbb.cam.ac.uk/news/2018JuneBIMBenefits (retrieved:14.10.2021)

NN (2019b) World Population Prospects 2019 – Highlight. https://population.un.org/wpp/Publications/Files/WPP2019_Highlights.pdf (retrieved: 14.10.2021)

NN (2019c) Only 11 Years Left to Prevent Irreversible Damage from Climate Change, Speakers Warn during General Assembly High-Level Meeting. https://www.un.org/press/en/2019/ga12131.doc.htm (retrieved: 14.10.2021)

NN (2020d) NBS's 10th Annual BIM Report 2020. https://www.thenbs.com/bim-report-2020 (retrieved: 14.10.2021)

NN (2021e) https://www.wbdg.org/bim/cobie/ (retrieved: 27.05.2021)

NN (2021ac) Construction Global Market Report 2021: COVID-19 Impact and Recovery to 2030. ResearchAndMarkets. https://www.globenewswire.com/en/news-release/2021/03/16/2193403/28124/en/Construction-Global-Market-Report-2021-COVID-19-Impact-and-Recovery-to-2030.html (retrieved: 14.10.2021)

NN (2021ad) BIM@SBB Road Map. https://company.sbb.ch/en/the-company/projects/national-projects/bim/documents.html (retrieved: 14.10.2021)

NN (2021ae) Forecast end-user spending on IoT solutions worldwide from 2017 to 2025. https://www.statista.com/statistics/976313/global-iot-market-size (retrieved: 14.10.2021)

NN (2021af) UK BIM Framework, www.ukbimframework.org (retrieved: 14.10.2021)

Patacas J, Dawoo, N, Kassem M. (2020) BIM for facilities management: A framework and a common data environment using open standards. Automation in Construction 120(December 2020). https://doi.org/10.1016/j.autcon.2020.103366 (retrieved: 14.10.2021)

Sacks R, Eastman C, Lee G, Teicholz P (2018) BIM Handbook. 3rd ed., John Wiley & Sons, Hoboken, New Jersey, 2018, 659 p

Sawhney A (2015) International BIM implementation guide – RICS guidance note, global. RICS, 1st edition. https://www.rics.org/uk/upholding-professional-standards/sector-standards/construction/international-bim-implementation-guide (retrieved: 14.10.2021)

Saxon R, Robinson K, Winfield M (2018) Going digital – A guide for construction, clients, building owners and their advisers. https://www.ukbimalliance.org/wp-content/uploads/2018/11/UKBIMA_Going-Digital_Reportl.pdf (retrieved: 14.10.2021)

Teicholz P (Ed.) (2013) BIM for Facility Managers. John Wiley & Sons, Inc., Hoboken, New Jersey, 2013

Thomas P (2017) The role of FM in BIM projects – Good practice guide. https://www.iwfm.org.uk/resource/the-role-of-fm-in-bim-projects.html (retrieved: 14.10.2021)

Wilson D (2018) Strategic Facility Management Framework – RICS guidance note, Global. RICS & IFMA, 1st edition. https://www.rics.org/globalassets/rics-website/media/upholding-professional-standards/sector-standards/real-estate/strategic-fm-framework-1st-edition-rics.pdf (retrieved: 14.10.2021)

Wright L, Davidson S (2020) How to tell the difference between a model and a digital twin. Adv. Model. and Simul. in Eng. Sci. 7(2020)13, 13 p

2 Digitalization Trends in Real Estate Management

Michael May, Thomas Bender, Joachim Hohmann, Erik Jaspers, Thomas Kalweit, Stefan Koch, Markus Krämer, Michael Marchionini, Maik Schlundt and Nino Turianskyj

Digitalization originally refers to the conversion of analog values into digital formats in order to be able to process such data using information technology. In the context of planning, construction and operation of built assets, this is mainly done today using BIM. The data can be considered a digital representation of the physical object and consists of the BIM model (i.e., the graphical element), together with alphanumeric information and documents which are usually stored and managed in one or more database(s). Of course, dealing with digital data requires new tools (IT systems), with which work is done, as well as new media for storage and new ways of data exchange, and importantly competencies to use such tools.

M. May (✉)
Deutscher Verband für Facility Management (GEFMA), Bonn, Germany
e-mail: michael.may@gefma.de

T. Bender
pit – cup GmbH, Heidelberg, Germany
e-mail: thomas.bender@pit.de

J. Hohmann
Technische Universität Kaiserslautern, Kaiserslautern, Germany
e-mail: joachim.hohmann@bauing.uni-kl.de

E. Jaspers
Planon B.V., Nijmegen, Netherlands
e-mail: erik.jaspers@planonsoftware.com

T. Kalweit
net-haus GmbH, Berlin, Germany
e-mail: t.kalweit@net-haus.com

M. May et al. (eds.), *BIM in Real Estate Operations*,
https://doi.org/10.1007/978-3-658-40830-5_2

However, through digitalization, not only does the format in which information is held and processed change, but also the processes and methods of work change. This results in completely new concepts and methods, which are generally referred to as digital transformation. In order to structure these changes and implement them sustainably, organizations, roles, responsibilities and processes must be redefined, and tasks coordinated. This also impacts the way in which industry professionals cooperate over all lifecycle phases—from cooperation to collaboration. In addition, this also creates new opportunities to develop and offer services for planners, developers, operators and users. Here it becomes clear once again that BIM is a holistic method that cannot be reduced to the mere existence of a digital building model.

The success of a BIM project therefore does not only lie in the digitalization (BIM model), but also in particular in the digital transformation (i.e. the information which is generated using BIM methods), with digital data being the basis of this digital transformation. Because the best model structure is of no use if the processes are unclear and nobody takes responsibility for entering important data into the model or ensuring it is kept up to date.

The following sections briefly explain relevant digitalization technologies and trends that are already playing an important role in the real estate and FM sectors today: Selected technical terms on the subject of digitalization in FM are also explained in the GEFMA glossary of FM terms (https://www.gefma.de/glossar/).

2.1 CAFM and IWMS

Facility Management (FM) is a management discipline that deals with the complex processes involved in managing and maintaining real estate. This includes all technical, infrastructure, operational services, planning and commercial tasks related to buildings

S. Koch
Axentris Informationssysteme GmbH, Berlin, Germany
e-mail: skoch@axentris.de

M. Krämer
Hochschule für Technik und Wirtschaft Berlin, Berlin, Germany
e-mail: markus.kraemer@htw-berlin.de

M. Marchionini
ReCoTech GmbH, Berlin, Germany
e-mail: marchionini@recotech.de

M. Schlundt
DKB Service GmbH, Berlin, Germany
e-mail: maik.schlundt@dkb-service.de

N. Turianskyj
LED-Studien GmbH, Panitzsch, Germany
e-mail: info@led-studien.de

and their facilities throughout their lifecycle. Real estate makes up a very significant part of the typical balance sheet of most organizations and assets are generally the second biggest cost after personnel costs. As such it is understandable that more and more attention is being paid to the professional management of real estate. Consequently, companies and public institutions are increasingly looking for cost-saving potential in the support (non-core business) areas of their organizations. The integration and sustainability of FM processes are also gaining importance. In the last two decades, it has increasingly been recognized that it is not possible to effectively and certainly not efficiently control the complexity of FM processes and the enormous volume of associated data without modern information technology (IT) and digitalization. This requires specific know-how and experience from successfully implemented digitalization projects.

For over three decades, Computer Aided Facility Management (CAFM) has been used to meet this challenge and a range of powerful CAFM software systems have been developed and used in practice.

CAFM software can be understood as an application software that comprehensively supports the digitalization of facility processes throughout the lifecycle of facilities. These processes, also referred to as CAFM core applications (NN 2021a), include, inter alia,

- Space management,
- Maintenance management,
- Inventory management,
- Cleaning management,
- Space and asset reservations,
- Lock management,
- Move management,
- Rental management,
- Energy controlling,
- Safety and occupational health,
- Help and service desk,
- Environmental protection management,
- Budget management and cost tracking,
- BIM data processing,
- Contract management and
- Workplace management.

The processing, evaluation and display of graphical and alphanumeric data to support these processes is just as important as the systematic control of workflow management and the integration possibilities with other IT systems (NN 2021a). CAFM can therefore be considered as the implementation and support of the FM concept using modern information and communication technology over the entire lifecycle of facilities.

In the English-speaking world, the term Integrated Workplace Management Systems (IWMS) has also established itself (cf. May and Williams 2017). The key components of IWMS are:

- Project management,
- Real estate portfolio and lease management,
- Space management,
- Maintenance and servicing (including CMMS—Computerized Maintenance Management System and EAM—Enterprise Asset Management) and
- Sustainability and operators responsibility.

This corresponds to the IWMS understanding of the term CAFM system in the German-speaking world and large parts in Europe.

In the literature (cf. May 2018a), the terms CAFM software and CAFM system are distinguished. CAFM software is a tool that can be purchased or licensed, rented or created by a user organization. A CAFM system, on the other hand, is an individualized and thus customized information system for the digitalization of facility processes. It provides the company-specific FM data. The distinction between software and system was made to make it clear to a CAFM interested party that, in addition to the purchase of a license or the rental of a CAFM software, a considerable effort is required for data provision and adaptation of the solution, before a running CAFM system is available.

A further discussion of the CAFM topic takes place in Sect. 3.2.

2.2 Building Information Modeling

Building Information Modeling (BIM) (Borrmann et al. 2018) describes a method in which all data generated during the design, construction and operation of buildings are brought together in a so-called Building Information Model which is maintained and used jointly by the participants across the lifecycle.

A wide range of stakeholders will use the same data model. This makes it possible to detect collisions when different specialists in the procees bring together their individual models (this is referred to then as a federated model) at an early stage. Redundant data is prevented so that all parties can communicate based on the same current state, thereby creating uniqueness and clarity. This enables effective communication and saves time and effort.

A building information model is understood to be a digital representation—a structured data set—of a building and related assets that either already exist or are in the planning stage. The structured data contains all the necessary information about the building (geometric and alphanumeric information). A building information model contains, in addition to the three-dimensional representation of the geometry, also semantic information such as costs, type information or technical properties (Borrmann et al. 2018).

Fig. 2.1 shows some typical attributes that can be included in a BIM model for components such as windows, doors, walls and floors.

The BIM model is usually the one and reliable source of information (Single Source of Truth) for a new building and its technical facilities (assets). However, it should be noted information may exist in other databases or sources outside the model. It is generated by suitable software tools, which are referred to in international language usage as *BIM authoring tools*.

A key feature of BIM authoring tools is object-oriented modeling. "Smart" components (objects) are modeled, containing all relevant geometric and alphanumeric information (component types and attributes) to describe the object.

In authoring tools, graphical and alphanumeric data are interlinked. A graphical object such as a wall (geometrically a cuboid) is enriched with attributes. Thus, the wall can be described as a component and it can be exactly determined which type of wall it is and what material it is made of.

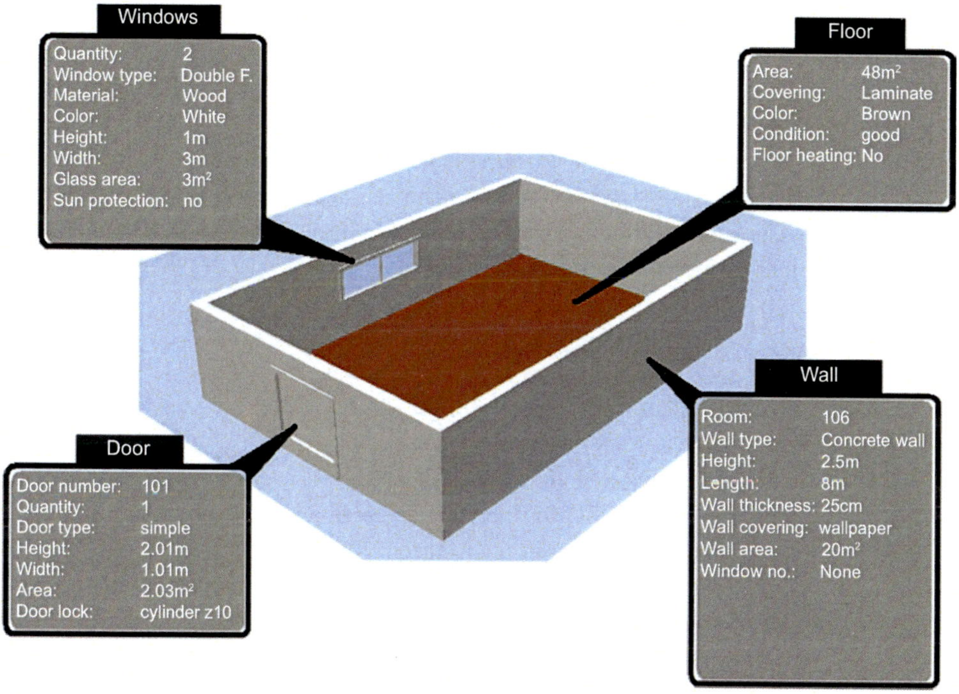

Fig. 2.1 Components in a BIM model with attributes

For further calculations, additional attributes can be stored, e.g. the heat transfer coefficient for energy calculations or also cost information for components, which can be used to determine the maintenance and running costs.

BIM and thus a central virtual building model with corresponding data result in numerous application cases (BIM use cases). Examples include supporting quantity take offs, cost planning, and energy considerations, and the possibility of taking operating costs into account as early as in the planning phase.

In addition, the building information model created in a BIM project provides a valid data basis for a building operation supported by CAFM (cf. Chap. 3 and Teicholz 2013).

2.3 IT Integration Technologies

A key aim of IT systems is to support business processes. However, with regard to their scope and complexity, business processes can vary widely. On the one hand, a business process requires the implementation of various work steps with specific business logic. On the other hand, it can involve one or more people who are active in one or more organizations, at one or more locations, and using one or more IT systems.

Against this background, the coupling of different IT systems is strongly influenced by the respective business processes. For this purpose, a wide range of different integration technologies are available, which are continuously developed further by the participating organizations in line with increasing requirements.

From the perspective of a system user, it is beneficial that only one system is used for the execution of a business process and to avoid the need to switch to another system. This can be achieved if System A—where required—provides files, alphanumeric data, certain functions or the logic or components of the user interface of other systems B, C, … (cf. Fig. 2.2).

2.3.1 File Integration

File integration can be achieved through an interface to a directory service (see Fig. 2.3). This should make it possible for files that are used and edited by many people and systems to be available at any time in their current form. In any case, it must be specified who made the last change to any file revision. A shared file storage can be set up via a central data storage with access via networks or using Internet protocols such as WebDAV. In a database, files can be mapped as large data sets in Binary Large Objects (BLOBs). In this case, access can be controlled via the database user concept and a database management system (DBMS). Numerous Cloud services offer external online storage. Some of these services allow multiple update access to files. This is usually the case with file-based cloud services that are connected to collaboration tools such as Microsoft Teams and Google Docs. If there is a need for file version control including check-in and check-out as well as release features when documents are updated, a document management system (DMS) can be integrated.

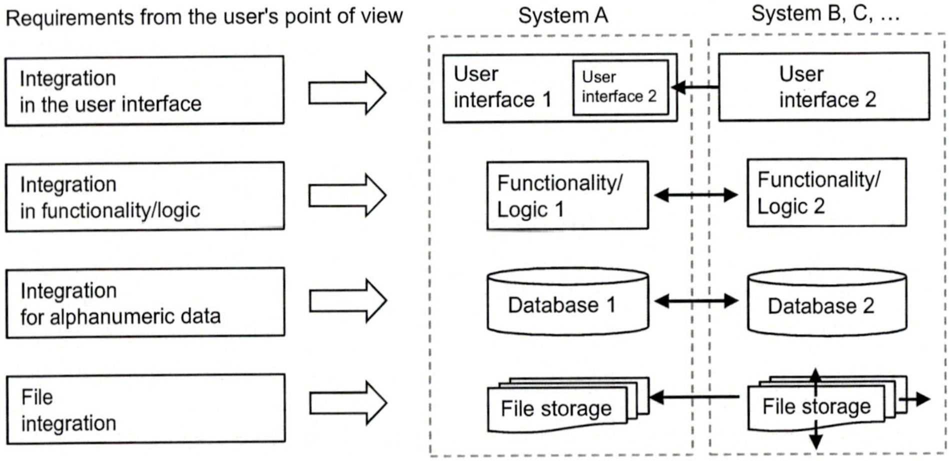

Fig. 2.2 Levels of integration depth from the user's perspective

Type of interface	Implementation	Online/Offline	Advantage	Disadvantage
Directory service	WebDAV, BLOB, Cloud services, DMS	Online	Current files available	Availability only with network connection
File interface	Export/Import Interfaces	Offline, Batch runs	Simple import of mass data	Individual, no Feedback
Database interface	SQL e.g. via ODBC, JDBC	Online, if necessary bidirectional	Quasi-standard for databases	Data structures Must be known
Functional Interface	Web service, API, RFC, RMI, Messaging, JSON	Online, if necessary bidirectional	Encapsulation of the Functionality	Limited Scope of functions

Fig. 2.3 Interface technologies for IT integration

Data is exported from one system and imported into another system via an exchange file (file transfer). The so-called batch run is unidirectional and asynchronous. This offline interface is widely used for non-time-critical requirements.

2.3.2 Integration of Alphanumeric Data

In order for systems to function properly, they usually require a minimum amount of data to be entered into the database. Data transfer interfaces are therefore widely used. Often these interfaces run at scheduled times. Since data structures of different systems are usually also different, data that comes from one system often has to be transformed before it can be loaded into the other system. In general, data transfers are carried out in three "steps" which are referred to as *Extraction, Transformation, Loading* (ETL). Extraction refers to reading the data from the source system, Transformation refers to restructuring the data into the format of the target system and Loading refers to inserting the data into the target system. ETL tools are available from various vendors. Some CAFM system vendors also offer them.

A database interface allows direct access to the respective database via standardized protocols such as Open Database Connectivity (ODBC) and Java Database Connectivity (JDBC) (see Fig. 2.3). This process can be used to create online interfaces.

2.3.3 Integration of Functionality and Logic

A functional interface uses the access methods provided and released by a software provider for the respective system. These interfaces encapsulate the databases against faulty accesses and allow the online call of functions or logics for use in another system. The current standard for functional interfaces is Webservices based on HTTP, of which REST (Representational State Transfer) has become very popular. Also used for this purpose are various Application Programming Interfaces (APIs), which are based on technologies such as Remote Function Call (RFC) or Remote Method Invocation (RMI) (see Fig. 2.3).

In addition, systems are available that are based on a microservices architecture. While traditional APIs use an agreement on actions that may be requested from a specific service, a microservices architecture uses message-based communication utilizing a messaging system to which messages are sent and which are then processed by microservices. The main advantage of this product category is that it can usually be integrated into other systems easily.

In addition to integration via webservices, direct message exchange is also gaining popularity. Here, systems create messages with predefined attributes and send them to other systems that process them. A typical component that uses this approach is called a *message bus* (Vaughan 2020). A message bus is a combination of a common data model, a common set of commands, and a message infrastructure that allows various systems to communicate over a common set of interfaces.

This approach has a number of advantages:

- *Logging*
  All messages exchanged between systems can be logged. This allows for complete transparency and traceability of system communication: the what and the when.
- *Queues*
  In many messaging systems, messages can be queued. This is particularly advantageous if the receiving system is under load and cannot process incoming messages immediately. This makes integration quite stable and scalable.
- *Secure*
  Most messaging services provide for a secure (encrypted) transmission of the data encapsulated in the messages.
- *Event-driven*
  Systems can set up and send messages that are directly related to user actions or other events that occur within the system. For example, if a user submits a repair request in a CAFM system A, this request can be sent by message to the system of the service provider, thus allowing for a quick response to the request.

2.3.4 Integration into the User Interface

Should there be a need for integration beyond documents, data and functions, this may require the user interface of one system or a component thereof is embedded in another system. In such cases an integration in the user interface takes place. For specific apps or viewers this can be achieved via HTML pages. The embedding of products from third parties is usually implemented using technologies such as JSON.

The Fig. 2.3 gives an overview of the interface technologies, possibilities for their implementation and the key advantages and disadvantages.

Further information on integration can be found in (May 2018a; NN 2014a, 2022c).

2.4 Mobile Computing

2.4.1 Mobility

In October 2016, a tipping point was reached where more people worldwide were using the Internet with Smartphones and Tablets than with traditional desktop computers (Lösel 2017). Although this growth was largely due to the development in African countries and Asian countries such as India and China, it clearly shows that the use of mobile hardware and software, as well as services based on them, has steadily increased and is still growing. Technical innovations with ever increasing service offers and an increasing demand are driving this development even further.

The term "mobile computing" includes mobile hardware and software, as well as mobile communication technologies including the protocols and standards on which these technologies are based. Typical device classes are notebooks, tablets and smartphones, smartcards and RFID devices, sensors and wearables, such as smartwatches or similar portable mobile devices.

The goal of mobile computing is, among other things, to provide a mobile user with supportive information and services based on his location and situation. Such services are referred to as location-based services. Mobile computing is characterized above all by three essential elements: mobility, networking and location.

"Mobility" in connection with mobile computing has three perspectives: (Bollmann and Zeppenfeld 2015)

- *Device mobility*
  A mobile device is networked with other infrastructure components regardless of time and place.
- *User mobility*
  Depending on the location and situation, the mobile user has the appropriate device available. For authentication, he uses unique security features such as passwords, PINs or chip cards.
- *Service mobility*
  A service is available regardless of time, location and hardware. A classic example is the use of an email service. This can be accessed and used on the go using smartphones, in the office using installed desktop software, and at home using a mobile web client.

2.4.2 Characteristics of Mobile Computing

Mobile Computing can be described in more detail using the following characteristics:

- *Portability/Resources of the devices used*
  Mobile devices generally use less local resources (processor, power supply, memory, display, etc.) than stationary devices during their work.
- *Connectivity/Properties of the connection*
  Mobile connectivity is variable regarding reliability and performance. Wireless connections usually offer a lower transmission performance than wired connections. In addition, this type of connectivity is very susceptible to external influences and can be greatly influenced by disruptions.
- *Security Requirements/Security aspects in mobile use*
  In addition to the security requirements that apply to stationary devices, mobile end devices and their infrastructure must also meet further security criteria. This area of data security is also referred to as Mobile Security. The fact that mobile devices are often located in environments that are difficult to control poses specific challenges for the operation of such devices. Also, the loss or theft as well as the connection via external access points can lead to an increased risk and must be considered within security concepts.
- *Usability/Use of mobile hardware*
  End devices in the field of mobile computing are often characterized by the fact that their design is adapted to the respective application and designed for complete mobile use. In addition, these devices are often tailored to specific applications or users. Multi-user devices, such as Unix- or Windows-based devices, are less common in this context.

2.4.3 Advantages of Mobile Computing

In contrast to stationary solutions, the decisive advantage of mobile computing lies in the continuous mobility and the permanent availability of mobile services, possibly depending on the communication infrastructure at the respective location (Lösel 2017).

In exchange for high local resource availability, the user receives flexibility in application usage as well as in functions that are only possible or optimized for mobile use. These include, for example, location-based services, GPS data processing for tracking assets and human activities applications or near-field capture using a camera, Bluetooth or Near Field Communication (NFC).

In addition, devices from mobile computing are increasingly focusing on alternative input methods such as biometric scans or voice control. This opens up new possibilities for people who cannot use traditional input methods due to a disability or local conditions. In terms of security, mobile devices and the installed technologies also offer advantages. For example, tracking and locatingfinding mobile devices is a helpful function.

2.4.4 Restrictions and Disadvantages

In addition to the advantages mentioned, mobile computing also has restrictions and disadvantages. For example, mobile devices are usually dependent on a built-in power supply in the form of battery packs. This limits the use more or less and has a significant impact on the design of the devices. The mobile power supply also poses challenging requirements on the installed technical components. Here, a balanced ratio between power consumption and possible performance must be achieved (Lösel 2017). However, the installed components and the small design partly restrict the usability. Displays, keyboards and other input devices are usually very small, so that alternative input methods, such as voice control, visual recordings (camera) and gesture recognition, must be used. This is sometimes still unfamiliar and requires practice.

Furthermore, the security aspect plays an important role in the use of mobile devices. Since these can be stolen or lost much more easily, more attention must be paid to their security. This is done, among other things, by increased protection of the devices using biometric login methods, PIN codes and similar techniques. The use of mobile devices in difficult-to-control environments, as well as the use of public access points, also increases the requirements for their security and can lead to restrictions in functionality. Another disadvantage is the use of mobile communication technologies, such as LTE or 5G. In contrast to the stationary environment, there can be considerable restrictions in a mobile environment regarding availability, reliability, range and quality in mobile use. Poor weather, difficult terrain or specific building materials, as well as the distance to the next access point, can reduce or even prevent signal reception.

2.4.5 Mobile Applications in the Real Estate Sector

2.4.5.1 Mobile Data Capture

Good quality and up to date inventory data is essential for optimal building operations. Capture and update of this data has been time-consuming and error-prone in the past. With the progress of digitalization in real estate and facility management, various systems for mobile data capture have emerged.

These systems provide the user with a variety of functions on site that allow data to be captured quickly and error-free. In addition to accessing already captured data, especially assistance systems, which support data capture, can be of great benefit. These include selection lists, plausibility checks, flexible search and filter criteria and data import functions that, based on the current location or similarity of objects, transfer information from any other data sources.

In general, mobile data capture systems in facility management offer possibilities for processing space and general building data, as well as data specific to technical assets (e.g. HVAC) and data for asset condition assessment. The most important advantages of using a system for mobile real estate data capture is ensuring high data quality as well as the minimization of the necessary data capture time. Data harmonization can be achieved directly during capture. Furthermore, effective control of the actual capture process is possible by continuous exchange of data between mobile devices and the central system.

Data obtained in this way can also be used in BIM models, e.g. assigned criteria or documents to technical equipment located in the BIM model which allows users to retrieve this information directly from the model.

Even though the availability and quality of mobile data connections are constantly increasing, there are still areas in the real estate sector where availability cannot be guaranteed (basements, technical rooms, etc.). Here, mobile data capture systems provide the user with functions that also allow offline work.

2.4.5.2 Mobile Document Management

In order to always have access to documents and files also on the move, various companies offer solutions for mobile document management. Mostly, these solutions are part of an holistic document management system and support, in addition to the access control, mainly enterprise-wide document-workflows. Documents are usually accessed via mobile devices, such as smartphones or tablets using native apps. Mobile access to document -workflows, such as the release of incoming invoices, is thereby just as possible as searching in the document base. Hereby, a central overview of search dialogues and lists is always guaranteed.

2.4.5.3 Mobile Field Services

In technical customer service and mobile field services it is essential to always have all relevant information about work orders directly at hand. In addition, access to all the customer's asset data as well as material stocks, can support the employees on site to carry out the necessary work efficiently and quickly. With functions, such as the work documentation, the capture of working time and material expenditure as well as documentating approvals by customers, mobile field service applications are able to digitally map and optimize what to date could only be done using paper-based activities. Therefore, these systems help to accelerate the flow of information faster and without media disruption. Costly re-entries of paper-based service reports or elaborate maintenance of information are thus no longer necessary. Especially to support real estate services, mobile field service applications are an important tool.

2.4.6 The Future of Mobile Computing

Mobile computing is today an integral part of the daily lives of many people and has also found its place in the business world. Especially the developments in the field of mobile workplaces highlight that mobile computing is gaining more and more importance. Supported by the growing number of cloud-based services and video conferencing systems, location-independent work is enabled almost everywhere and at any time. Furthermore, the new technologies in the field of mobile computing and fast expansion of mobile infrastructures are enabling new business models in B2B and B2C environments, which also promote the use of mobile technologies.

The potentials of mobile computing were recognized very early on in facility management (Hanhart 2008). This is also reflected in the observation that for years the use of mobile technologies in CAFM has been named as one of the most important, often even as the most important trend of software users and providers in the CAFM Trend Report of GEFMA (NN 2021k). It can be expected that the trend towards the use of mobile hardware and software will increase considerably throughout the real estate industry. Mobile access to BIM models or digital twins (cf. Sect. 4.1), in combination with augmented reality techniques (cf. Sect. 2.6), are empowering new business models as well as better and faster decisions in planning, construction and operation processes.

2.5 Cloud Computing

Cloud Computing has developed from an IT hype to a common and accepted technology in the last two decades and cloud computing has complemented previously used IT operating concepts in a meaningful and innovative way. Cloud computing offers companies with less strong IT skills the opportunity to react flexibly to changes in operational business and to avoid excessive or risky investments (NN 2016b).

Depending on the field of application, there are different definitions for the term "cloud" or "cloud computing". All of them have in common that they focus on the use of various IT resources via the Internet. The Federal Office for Information Security (BSI) also comes to a similar explanation, defining cloud computing as follows:

> Cloud computing is a model that allows access to a shared pool of configurable computer resources (e.g. networks, servers, storage systems, applications and services) on demand, anytime, anywhere over the Internet. They can be provided quickly and with minimal management effort or interaction with the service provider. (NN 2021i)

Other approaches deviate from this and rather focus on billing for used services.

However, cloud computing is generally described by five essential features (NN 2021i):

- *On-demand Self Service*
  Provisioning of resources (e.g. computing power, storage) takes place automatically without interaction with the service provider.
- *Broad Network Access*
  The services are available over the network and accessed through standard mechanisms and are not bound to specific client software or devices.
- *Resource Pooling*
  The provider's resources are in a pool which many consumers can access (multi-tenant model). The users do not know where the resources are located, but may be able to specify the storage location, e.g. region, country or data center.
- *Rapid Elasticity*
  The services can be provisioned and released quickly and elastically, in some cases even automatically. From the user's point of view, the resources therefore appear to be unlimited.
- *Measured Services*
  Resource usage can be controlled and optimized by leveraging a metering capability and made available to cloud users accordingly.

In addition, the different types of cloud computing can be classified according to their technical features. They are also generally identified by the addition "as a service" in the name. This classification is often depicted in a pyramid representation (see Fig. 2.4).

This illustrates that the four different manifestations of *Infrastructure as a Service* (IaaS), *Platform as a Service* (PaaS), *Software as a Service* (SaaS) and *Business Process as a Service* (BPaaS) build on each other (NN 2009).

IaaS is the use of shared capacity directly from the provided system via a network (e.g. memory or computing power). Here, the cloud system automatically adapts the resources made available to a user to their use. Technologically, this is achieved by virtualization of the resources, so that logically provided resources can be separated from the hardware basis and made available according to need.

A higher level than the mere use of resources is PaaS. Here, an infrastructure is provided that can meet certain requirements, such as databases and multi-tenancy. On this platform, the user can operate their own applications.

SaaS is probably the best known and most used manifestation of cloud computing. However, there are very different interpretations of what SaaS provides. In general, however, SaaS describes the provision of software applications over the Internet. This includes the provision, maintenance and administration of this environment.

BPaaS is not described by the BSI, however other sources describe the mapping of business processes based on cloud technologies as the fourth level of classification. BPaaS also offers the possibility to define processes, to monitor their course and finally to automate their execution.

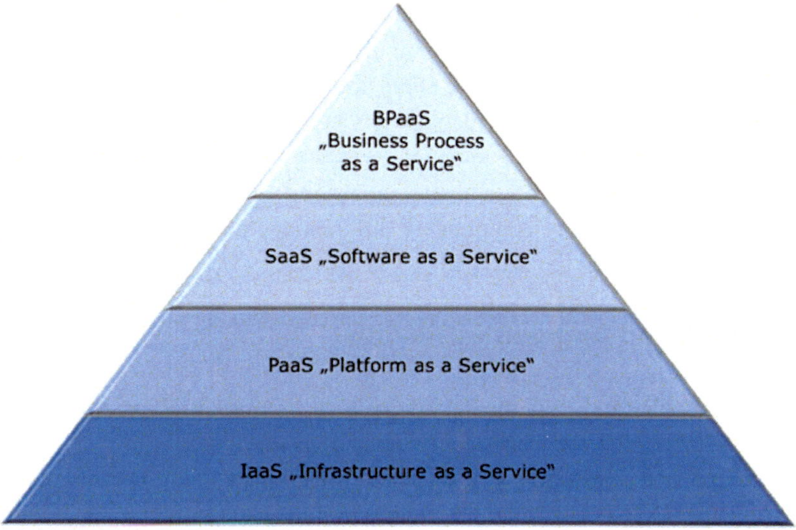

Fig. 2.4 Pyramid model of cloud computing

In addition to the technical classification, cloud computing can also be divided into four organizational concepts according to the National Institute of Standards and Technology (NIST), as Mell and Grance 2011 show:

- *Private Cloud*
  A private cloud is a cloud infrastructure that is exclusively provided for one institution. It can be organized and operated by the institution itself or by a third party and can be located in the data center of the institution's own or a third party.
- *Public Cloud*
  If the services are provided to the general public or a large group, such as an entire industry, by a provider, this is referred to as a public cloud.
- *Community Cloud*
  A community cloud is an infrastructure that is shared by several institutions that have similar interests. Such a cloud can be operated by one of these institutions or by a third party.
- *Hybrid Cloud*
  If several independent cloud infrastructures are used together via standardized interfaces, this is referred to as a hybrid cloud.

In general, cloud-based solutions are characterized above all by their fast availability and an easily scalable and accessible infrastructure. The high scalability makes it possible to adapt to the current business needs of an organization and its future development at any time, thus avoiding bottlenecks or overcapacities. There are also advantages from an economic point of view. For example, high investment costs for software and hardware purchases are eliminated. However, the degree of individualization within cloud solutions is rather limited. In addition, when using cloud solutions, it must be checked to what extent a transition to other software providers or service providers is possible. Especially with SaaS there may be restrictions.

However, cloud solutions offer a very interesting alternative to traditional operating models and enable innovative solutions for different application cases.

2.6 Mixed and Augmented Reality

Mixed Reality (MR) and in particular Augmented Reality (AR) are among the digitalization technologies that are increasingly being used successfully in real estate and FM. Many potential uses can only be insufficiently exploited today because there are still major problems in providing the often required information on site in the building. Especially in FM, there are many activities that must be carried out directly in the building or at a facility. For example, maintenance management may be mentioned which often requires knowledge about the real estate portfolio of an organization, including the technical assets and systems of the buildings. In order for the maintenance to be efficiently and effectively carried out information must be available in real time on site. Very often, diagrams, pictures, instructions and other documents are not available or not up to date, leading to wasted and avoidable effort and costs and, in the worst case, to wrong decisions. This in turn can lead to delays, financial losses, technical damage or even health risks.

By developing mobile devices which can provide a clear presentation of information and associated software as well as wireless communication, typical FM tasks can be handled faster, safer and more economically.

It is particularly helpful if the required information such as texts, pictures or models can be compared with the real situation and supplemented or corrected if necessary. This is increasingly done by overlaying the real objects with virtual models, for which suitable viewing devices such as tablets, smart glasses or head-mounted displays (HMD) can be used.

Fig. 2.5 shows the extension of reality by superimposing location-based virtual information (3D scenes and additional information) in an HMD.

Mixed Reality comprises such situations and systems in which human perception is superimposed or mixed with computer-generated perceptions. The wide range of applications is located between the real world (reality) and a purely virtual world (virtuality). Reference is made to Milgram et al. (1994), who postulated the so-called “reality-virtuality” continuum (cf. Fig. 2.6) with a continuous transition between the real and virtual environment.

Most of the MR applications interesting for the real estate industry can be assigned to augmented reality. While the use of purely virtual applications such as a virtual tour (walkthrough) in building design has been standard for many years, AR has only slowly

Fig. 2.5 Use of HMDs to enrich reality with virtual information

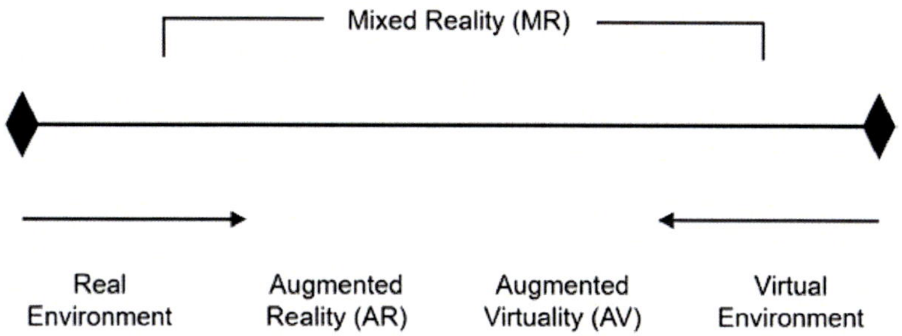

Fig. 2.6 Milgram's Reality-Virtuality Continuum

entered everyday life in the real estate industry in the last 10 years, which is due in particular to the technical requirements that must be met for AR.

Augmented reality exists when reality is interactively enriched with virtual content in real time, which is positioned at specific reference points in reality. AR thus enables the integration of virtual content into the real three-dimensional world (Ellmer and Salzmann 2014; May et al. 2017).

With AR, virtual information can be integrated into the real field of view in real time, thus, creating an informative enriched 3D scenario. Users of this technology perceive virtual objects as an additional part of the real world and can interact with them. This way of human-machine interaction, which makes computer-generated content experienceable in space, is made possible by the rapid development of mobile devices.

The challenges in AR technology are interaction in real time, recognition and tracking of real objects as well as positioning of virtual objects in the image of the real environment.

Basically, an AR device must provide a display or projector to enable visual enhancement of the real world. Mobile data connections via a mobile network and WiFi connections are required on the one hand for synchronizing with centrally stored data and on the other hand for determining the position of the AR device. One of the most important foundations for an AR application is the accurate spatial reference between the real world and virtual elements. GPS positioning alone cannot be used due to insufficient accuracy and often lack of satellite connection in buildings. Hybrid methods that access appropriate systems depending on availability and reduce the necessary computing power are most promising. The position of the AR device is usually determined in two phases. In the first phase, initialization, a precise determination of the position must be made using optical tracking by comparing it with stored point clouds or special markers. In the second phase, GPS and WiFi Positioning Systems (WPS) can be dispensed with, and optical tracking can be reduced to a resource-saving algorithm for tracking optical flow. In addition, sensors for rotational and translational movements as well as the Earth's magnetic field can be used as support.

New possibilities for AR arise when the spatial-geometric data and possibly factual data can be provided by a BIM model. One can then also speak of BIM-based AR (May 2017). Fig. 2.7 (cf. May 2018a) shows a tablet with the BIM model of a technical facility, which overlays the real asset and thus provides additional technical and visual information.

2.7 Big Data and Analytics

In this section we discuss the topic of data and its analysis in connection with the use of BIM and in combination with FM data. Hence, it focuses on the use of data and BIM during the operational phase.

BIM-related data analysis is widely discussed today and is closely related to the topic of the digital twin dealt with in Sect. 4.1.

Fig. 2.7 BIM model with highlighted component, which overlays the real asset

2.7.1 Relevant Data Classes

Any structural analytics service relies on trusted and structured data sets that are available to it. In connection with the topic of Big Data Analytics and in the context of BIM, the relevant data classes that can play a role in the analysis of buildings and physical workplaces must be identified.

2.7.1.1 BIM Geometry Data

A unique character of BIM models is that they provide geometric/spatial information. These *geometry data* differ fundamentally from other data types with which one usually works, such as process data, financial data and asset behavioral data, which are usually referred to as *IoT data*.

Geometry data can be analyzed to determine geometric information such as distances between objects and spatial angles between objects. This allows the position of interrelated objects to be determined in 3D space. Geometric data in particular allow the calculation of surface areas and volumes of building elements, which are important basic parameters for operating buildings.

The problem, however, is that BIM geometry data captured in BIM modeling tools is often represented in the proprietary format of the authoring tool. With using IFC classes a unification option of the BIM geometry data is offered. However, there are some specific limitations to the use of IFC in practice. One is that not all BIM authoring tools support a complete transformation of their internal format to IFC. Another limitation is that IFC models are designed to be non-editable. The original purpose of IFC was to allow for clash detection between various models coming from different BIM authoring tools during design.

In the design and construction phase, this aspect of IFC classes is not a big problem, because the project is time-bound and will end once the building is constructed. However, in the operational phase of buildings, the shape of buildings regularly changes through renovations, resulting in regular updates of geometric and alphanumeric data. This process of change needs to be managed well for analytics to remain valuable.

BIM viewers today represent the most important tool for displaying geometry data of buildings. There are hardly any other toolsets apart from GIS systems with the capability to represent location and geometry. BIM-related analytics will typically involve a model viewer, associating other data and information to the geometry representing the building.

2.7.1.2 BIM Asset and Element Data

This data represents the typical alphanumeric side of the model, describing the properties of assets and elements as modeled. This data is typically contained in a file-based format with a structure that is usually proprietary to the BIM authoring tool used. Converting the model to IFC will bring the data structure to a standard but again, sometimes incomplete. But in IFC too, the asset and element data will remain in a file format still. The structuring is often done in a hierarchical document or tree structure (see Fig. 2.8). This represents a fundamental limitation to analytics, especially in the big data context.

To overcome this impediment, many initiatives on creating so-called CDEs (Common Data Environments) are emerging. These are repositories of BIM related data which provide APIs (Application Programming Interfaces) to retrieve data from them, usable for analytics. The CDE technology is still in its early stage of development. Today, the most used approach is to extract the asset and element data from the model and store it in a differently structured data base, allowing for the data to be included in analytics processes (Sect. 4.3).

The most common transformation used here is to CAFM/IWMS systems, in which this data is mostly represented in a relational structure, quite useable for analytics. Fig. 2.9 shows a corresponding entity-relationship model.

2.7.1.3 Behavioral Data

Buildings and building components can provide data on their operation and condition. Today, sensors of all kinds are retrofitted to make existing buildings and their assets smart and connected, often using IoT technologies. In most cases, the parameters measured are stored in time-sequenced (historized) fashion, e.g. per hour or quarter. As a result, a so-called time series data set emerges (see Fig. 2.10).

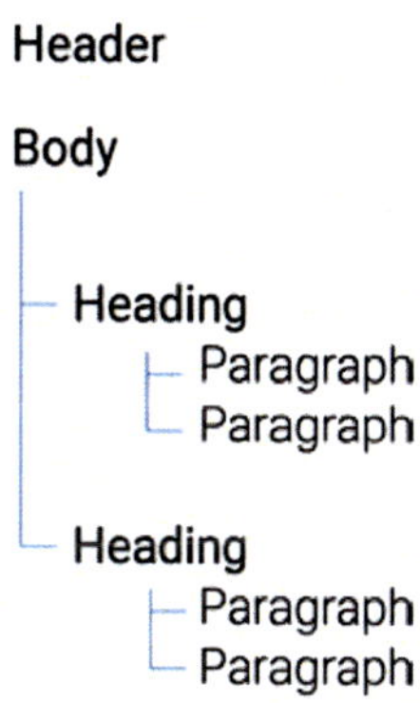

Fig. 2.8 Hierarchical data structure as a tree

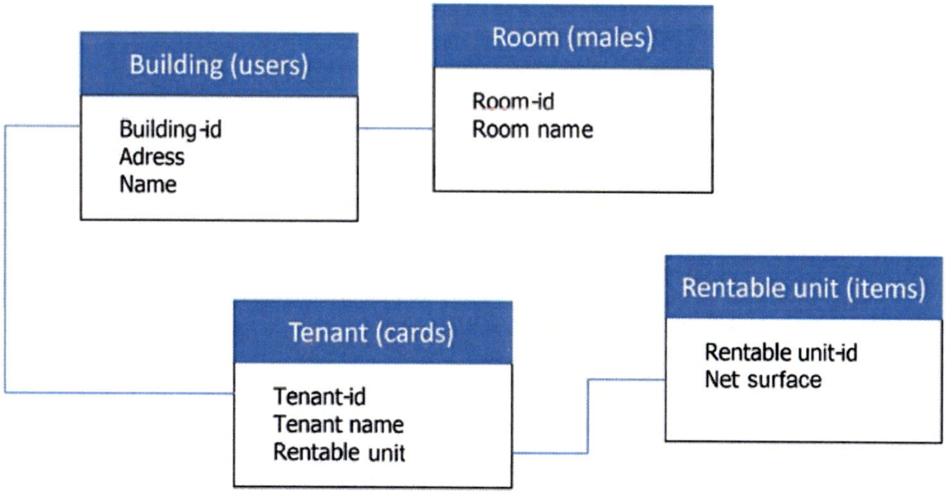

Fig. 2.9 Data structuring as a relational model

Time Series ID	Time-window start time	Temperature	Carbon dioxide	Humidity
101	15:50:00	20.0	800.5	65
101	16:00:00	21.3	910.0	67
102	16:10:00	22.5	1100.3	82
103	16:20:00	20.5	850.7	70

Row key

Fig. 2.10 Time series data set

This typically results in large amounts of data (big data) over time. To facilitate (real-time) analytics, a new type of data storage is being developed, which is referred to as time series data storage systems and optimizes the use of corresponding database management systems.

However, when data volumes are manageable and the analytics requires no real-time capability, the storage in well-designed relational databases occurs as well.

2.7.1.4 Business, Process and Financial Data

This data refers to the typical data sets that are being managed within CAFM/IWMS systems. Here, the data volume, which describes all business processes, is usually manageable. CAFM/IWMS systems are designed to support day-to-day operations and allow for high volume of data transactions (including changes to the data) to take place fast.

To provide the required performance, typically the relational data structures of this kind of data are designed for transaction speed. They are also referred to as *normalized* data structures. Their design is less favorable for analysis because these structures are often complex and difficult to interpret.

To eliminate this disadvantage, alternative data structures are sometimes used for analysis. In this type of so-called *Business Data Lakes* the data of objects is combined in simplified data table structures where data attributes may well be redundant (see Fig. 2.11). The data records are used only for reporting and analysis, not for change.

2.7.2 Analysis Options and Analytics Options

What types of analysis are feasible in view of these data sets available? The scope for analytics is restricted in this section to explicit relations with the geometry data. This implies that in the majority of cases, a model viewer is engaged in the analytics out-

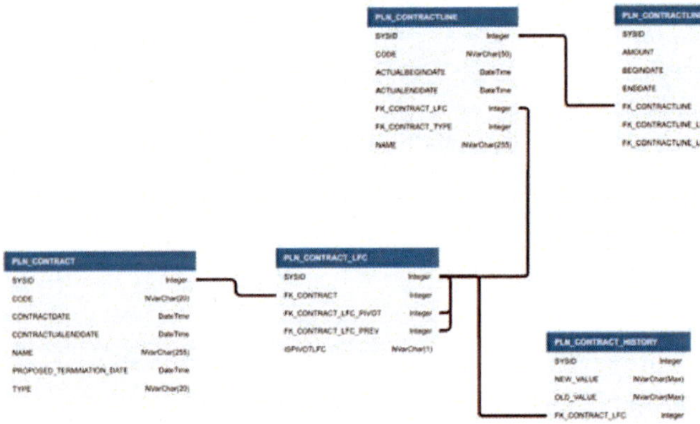

Fig. 2.11 Intentional redundancy in the data model of a data lake

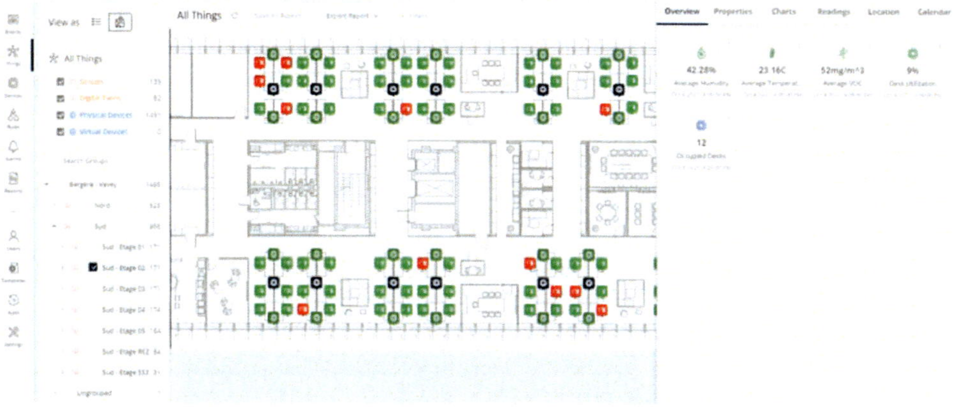

Fig. 2.12 Visualization of IoT data in a geometric model

put. Fig. 2.12 and Fig. 2.13 show IoT data in a geometric model and various workplace parameters like temperature, humidity, carbon dioxide, airborne chemicals and desk occupancy rate over the course of time.

The *analysis* in most BIM-based examples revolves around business data and behavioral data in the geometric context of the building. This combination provides an intuitive understanding of the respective situation. Time-series data is often represented in models using color coded schemes that indicate actual parameter values, where the user can specify the time to be displayed. As he moves through the time series, the image changes and provides insights into the dynamics.

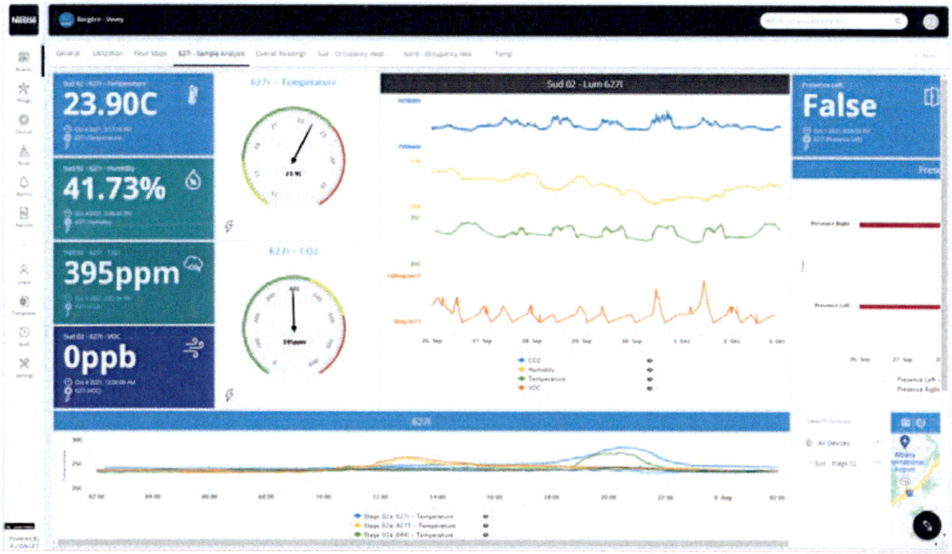

Fig. 2.13 Workplace parameters monitored over the course of a week

Many combinations of analyzing data sets are feasible in this context. The analysis of time-series data may disclose eminent failure on asset level, calling for repair or replacement. Combining this with geometric data provides good insight into the impact of the work at hand.

Today, a variety of analytics applications are available in combination with BIM models. The challenge is to connect all data sources in such a way that they can be analyzed at all. There will still be enough challenges to be overcome. A practical approach to overcome relevant challenges can be to check the BIM capabilities of CAFM/IWMS providers and their systems in this area and use them.

2.8 Internet of Things

Anyone who deals with trends in the real estate industry cannot avoid terms such as Internet of Things (IoT), Smart Home, Smart Building or Digital Twin. They all have in common that buildings, regardless of their use for living or business, must increasingly orient themselves to the needs of their users.

The technological basis for smart or digital buildings is a suitable technical infrastructure. Well established solutions are Building Management Systems (BMS) or Building Automation Systems (BAS). These solutions often include sensors for data capture as well as actuators for active controlling the installed technical facilities. However, these actions usually take place in a closed (proprietary) system of a manufacturer or by means of separate technical implementations (bus systems).

The internet of things goes beyond here. In (NN 2021g) IoT is defined as

> a collective term for technologies of a global infrastructure of information societies that allows physical and virtual objects to be interconnected and to cooperate with each other through information and communication technologies.

It is described in more detail in (NN 2021h):

> "The internet of things, or IoT, is a system of interrelated computing devices, mechanical and digital machines, objects, animals or people that are provided with unique identifiers (UIDs) and the ability to transfer data over a network without requiring human-to-human or human-to-computer interaction."

All relevant, uniquely identifiable objects (things) should be interconnected in an information technology structure. For the real estate industry, the following objects can be listed as examples:

- Building parts with condition information (rooms, windows, doors, …),
- Objects in the outdoor area with condition information (parking spaces, access, …),
- Technical facilities in the building with condition information and control options (heating, ventilation, air conditioning, elevators, shading, …),

- Technical facilities in the outdoor area with condition information and control options (barriers, security and access technology, …) or
- Users of the property (technically represented by their smartphone, their smartwatch or their vehicle).

Technically, IoT can be implemented with different approaches (also in mixed operation mode). In newly constructed buildings or if an existing building management system is already installed, communication can take place via wired systems. In existing buildings, where retrofitting cabling is not possible or uneconomical, wireless systems can be used. For technical details, see (May 2018a; NN 2022a).

In contrast to BMS, significantly more data and information about the state of assets are usually captured by IoT. In order for these data to be really used and not become a data graveyard, they must be prepared and processed. This is where the benefit of IoT begins. By processing the sensor data and combining different information, the current state can be influenced by algorithms such as machine learning. The insights gained help to increase the comfort of users, but also to reduce building costs. This is also due to the external data that does not belong to IoT according to the definition, such as weather and climate data, price information for media (e.g. electricity and water), but also usage data of the property.

The representation and use of all this information can be done using the concept of the digital twin of a building. The basis for this is the spatial assignment of the IoT sensors in the BIM data model and the possibility to retrieve the sensor data via the BIM model. For more information, see Sect. 4.1.

A rough classification of the topic IoT in a BIM/FM environment is shown in Fig. 2.14. Among other things, an architectural BIM model is generated by the authoring software. In parallel, sensor data is captured on an IoT platform, aggregated and possibly prepared and analyzed for visualization. The real added value only arises when the IoT data are linked to the data of the architectural model and the current process data from the CAFM software to form a digital twin.

IoT based on a digital building model and in combination with external data can significantly increase the comfort of users in the building and reduce operating costs of the building.

2.9 Artificial Intelligence and Machine Learning

Artificial Intelligence (AI) is a traditional field of computer science with significant benefits but also risks—this also applies to the use of AI in real estate and FM (May 2018b). AI will influence our lives and thus our working environment to a greater extent than any other technology in recent decades.

The basics of AI go back to the 17th century, when Charles Babbage had the revolutionary idea of an *Analytical Engine*. But it was not until the 20th century that the condi-

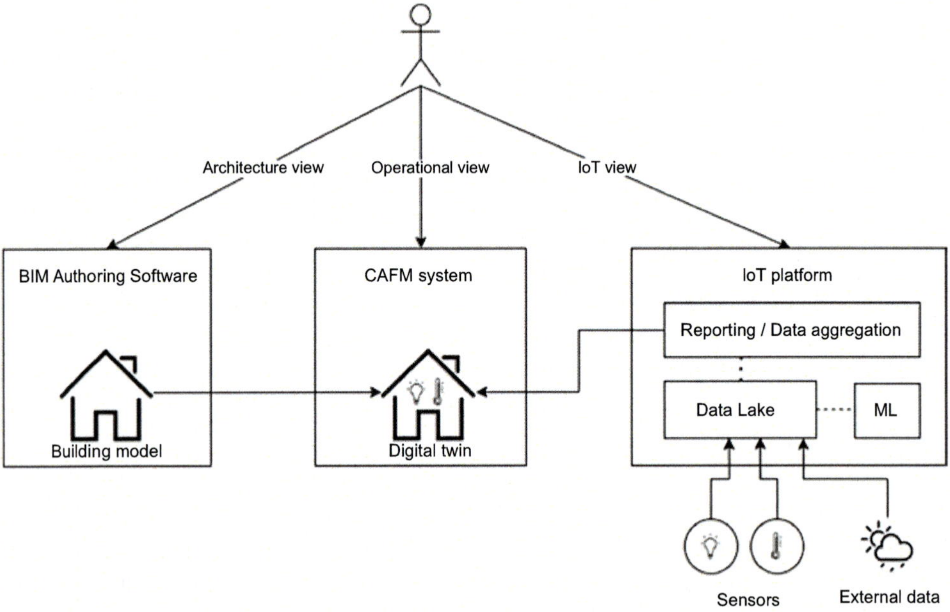

Fig. 2.14 Role of an IoT platform and digital twin in a BIM/FM environment

tions were created to not only conceive of a functional (electronic) computer, but also to build one (e.g. by the German computer pioneer Konrad Zuse).

We are still far from being able to finally define the term *Intelligence*. AI tries to imitate human perceptions as well as human decision-making and actions through machines. AI is a field of computer science that is dedicated to solving cognitive problems that are often associated with human intelligence, such as learning, problem solving, argumentation and pattern recognition. Some features that intelligent behavior must at least have are undisputed (Schneider 2012). These include learning ability, the ability to make logical inferences, planning ability, problem-solving ability and motor intelligence.

Many disciplines work together for this. In the beginning, these were the theories of axiomatic reasoning, mechanical calculations and the psychology of intelligence (Hofstadter 1985). Today, additional fields such as cognitive science, neurology, evolution, statistics, multimedia analysis and data mining, linguistics and even philosophy have been added. After decades of intensive research, we had to realize that a "thinking" machine cannot be constructed without intensively researching human thought.

The boundaries between intelligent and non-intelligent behavior are still not clearly defined. It is undisputed that intelligence at least includes learning, creativity, emotional reaction, sense of aesthetics and self-consciousness. But when is a machine considered to be intelligent? This question has occupied AI researchers for many years. The generally accepted measuring instrument is the so-called Turing test from the year 1950, named after the famous English mathematician Alan Turing, who called this test the *Imitation Game*.

A human communicates parallel with another human and a machine without visual or auditory contact, e.g. via a keyboard and a screen. Both conversation partners (human and machine) answer questions and try to convince the questioner that they are thinking people. If the tester cannot clearly decide after the conversation which of the two conversation partners is the machine, the machine has passed the test and may be considered intelligent. So far, hardely any computer program is known to have passed the Turing test. Google's LaMDA or chatbots like Goostman or ChatGPT claim to have passed the Turing test. But this is still up for debate among the scientific community. This may be an indication of the complexity of natural intelligence. However, we are already approaching the limit with today's chatbots where we can hardly decide without further ado whether we are dealing with a human or a machine. ChatGPT is an impressive recent example of what AI can do in the area of text generation.

In the past, AI has been able to celebrate successes in certain limited task areas. These include board games such as chess and the much more complex game of Go, the use of robots in manufacturing and healthcare, and also in finding mathematical proofs. Meanwhile, however, the scientists are much more ambitious. This has resulted in ambitious plans to build a so-called general AI—that is, a system that not only performs a clearly defined task, but also comprehensively understands the world, can orient itself in it and can solve any problem (Göring 2017). This is about nothing less than developing machines that are just as intelligent as humans or even more intelligent. For many researchers, it is obvious that today's AI already has forms of consciousness. So they are curious, creative and show individuality. However, predictions about the rapid development of intelligent machines had to be corrected again and again during development (Buxmann and Schmidt 2019).

In addition to the many existing and common model types such as data models, statistical models, role-based expert system models, operations research models, the AI represents a new class of models. AI models are unique in that they can learn. For this they use extensive data that they are "fed" over time again and again. As a result, they are usually able to improve their perceptions and reactions over time.

Machine Learning (ML) is a sub-discipline of AI that refers to methods and technologies that are used for learning. ML is about enabling machines to learn independently. Often patterns must be recognized and predictions made. Learning can take place either trained or uncontrolled.

A typical example of the use of ML and pattern recognition is autonomous driving in which dangerous situations must be recognized and assessed in real time and immediate action must be taken (cf. Fig. 2.15).

The application of machine learning requires specific competencies. This applies not only in terms of data management, but also in terms of designing and configuring neural networks as well as in the learning phases during which data is fed into the network and corrections are made based on the results. As a rule, very large amounts of data (big data)—often more than a hundred thousand data sets—are needed for a neural network to learn effectively. Machine learning can work with very different data types. A well-known example of this is the ability of these systems to analyze images and describe what can be seen there. Today, computers recognize faces with better accuracy and reliability than humans.

Fig. 2.15 Use of AI in autonomous driving

Many advances in AI are based on artificial neural networks (ANN). These include hardware and software methods that attempt to mimic the human nervous system. Information processing takes place in a similar way to nature, with information being passed on through connections between the (artificial) neurons. Although the first studies date back three quarters of a century, real interest only arose in the mid-1980s when it was discovered that certain complex optimization problems could be solved using ANN and improved learning methods (e.g. backpropagation) were developed. ANN often create (but not exclusively) the basis for the different forms of ML that are increasingly gaining importance in the real estate industry. For complex ANN, this is usually done via so-called deep learning methods. In addition to commercial development platforms for ML applications, there are now also numerous open-source frameworks available (see Hwang 2017).

A good overview of application areas for AI in the fields of design and construction, real estate and smart cities as well as FM is given by Hoar et al. (2017) (see also Altmannshofer 2018; May 2018a). In the field of facility services, for example, catering, reception/helpdesk, cleaning, security, inspection, maintenance, space occupancy and -management as well as logistics are mentioned. But there are also examples of AI in generative design (May 2020). Here, too, the disruptive character of AI is expressly pointed out with the request that it is high time to also build the necessary competencies in the real estate sector in order to use this technology profitably. But ethical questions also need to be addressed.

In the real estate sector, ML is already being used today for predictive maintenance (May 2018a). For this purpose, data is used that describe events and defined parameters over time for a technical asset, the behavior of which can usually be recorded using a quite limited set of parameters and whose behavioral variance is limited.

However, it should be noted that the application of ML for the prediction of events becomes quite complex when the number of variables involved in the behavior to be described is high and the behavior to be predicted can vary significantly.

2.10 Digital Workplace

2.10.1 Digital Workplace in CAFM/IWMS

In general, "Digital Workplace" refers to a networked and smart working environment that allows companies and organizations to work together to a large extent independently of location and time based on digital technologies with corresponding tools and services. In this context, in connection with CAFM/IWMS, digital workplace refers to the functionality of providing and managing workplaces equipped with modern IT, including the associated services. Not only classical office workplaces are considered, but also generally those workplaces that have a similar organization and technical equipment (e.g. in a warehouse or in production).

Supported sub-processes and process steps

The following sub-processes and process steps for the design, management and operation of digital workplaces with their required CAFM functions are described below:

- Requirements analysis for the number and equipment of workplaces,
- Establishment and provision of workplaces and
- Booking, occupancy and billing of workplaces.

For the terms *occupancy* and *booking* used here and below, the following understanding applies: The term occupancy refers to a permanent use of workplaces during a certain period of time and the term booking to a short-term use. In this way, in an occupancy planning, for example, shared desk workplaces with their equipment can be "occupied" by certain departments over a longer period of time, which in turn can or must be booked in a specific, limited use.

Functions

The following functions are to be provided for process support by CAFM/IWMS software:

- Mapping of a structural element or object "workplace" (physical workplace) with linking to rooms or room zones,
- Assigning attributes for equipment of this object,
- Assigning a maximum number of placeable workplaces of a certain type per room or room zone based on an area standard,
- Differentiation of territorial and non-territorial workplaces,
- Displaying workplaces in a floor plan (horizontal) or a stack plan (vertical) with symbols,
- Displaying reservations/bookings,
- Displaying occupancy,
- Linking workplaces with their "occupancy",
- Entering a temporary occupancy, also as a prerequisite for a workplace booking system,
- Booking of workplaces, inter alia with
  - Booking according to temporal, price, equipment-oriented features,
  - Group bookings of workplaces taking into account relations for interaction and collaboration,
  - Booking of workplaces via a (also external, e.g. Outlook) calendar function individually or in series,
  - Booking of additional services,
  - Storing of guidelines for booking of workplaces,
- Occupancy planning of variants and in considering time-dependencies with the aim of space compaction and/or localization, e.g. of new employees,
- Derivation of subsequent moves after confirmation of a variant and
- Definition of complete workplaces as moving objects including information such as equipment features, occupancy and booking rules.

Data and catalogs
The following data and catalogs are to be processed primarily:

- Desk type X, docking station type Y, sensor type Z,
- Distinction between fixed assigned and freely assignable workplaces,
- Equipment symbols in the floor plan,
- In case of permanent use with date from—to; in case of temporary booking day/time from—to,
- Workplace occupancy (also anonymous) and
- Cost calculation.

Reports and evaluations
The following reports and evaluations are to support the decisions in Workplace Management:

- Current occupancy or booking of individual or a group of workplaces,
- Connection to Workplace Monitoring by sensor/IoT/IP addresses via a defined interface,
- Analysis of potentials for occupying/utilizing workplaces based on key figures and graphics,
- (Time-related) statistics on workplace utilization,
- Evaluation of the use of workplaces per billing unit and
- Costs of workplaces for calculating and billing of usage.

Interfaces
Interfaces are to be provided, inter alia, to:

- IoT platforms for determining the actual occupancy and controlling the comfort or compliance with minimum distances,
- Calendar systems as part of office software or as an app on a smartphone or tablet, and
- ERP systems for billing the use of the workplaces.

2.10.2 Digital Workplace Management Systems

The experience of hybrid working models that was imposed on us in a short period of time due to the COVID-19 pandemic has led to a widespread use of specific stand-alone digital workplace management systems (WMS) in terms of their variety of offerings and practical applications. In summer 2021, the authors had already identified more than 50 such systems on the German market, which are frequently offered by PropTech companies (see NN 2021aj), but also by long-established software suppliers. The functionality of workplace management systems is comparable to that available in CAFM/IWMS modules (see Sect. 2.1).

Conceptually, workplace management systems differ from many older CAFM/IWMS systems, especially in the following respects:

- 100% cloud-based,
- Use of APIs,
- Only minimal customizing possible and necessary,
- Short introduction times,
- Low initial costs,
- Only usage costs (pay per use) and easily scalable pricing models (quantity-based, basic, extended or premium package),
- Inclusion of all types and equipment of workstations, including home office and parking,

- Access via app, portal or kiosk,
- Networking with access control systems and
- BIM integration (selective).

Example of a cloud-based app

The following example of a cloud-based app is the software Seedit (NN 2021ak). The app is an organizational control tool that primarily creates transparency about the use of company facilities through a booking system. Fig. 2.16 shows the two essential components of the system with users and workplaces.

The use and booking of the following resources can be displayed transparently by:

- Workplace and home office for individuals or teams,
- Conference rooms including catering,
- Zones for confidential meetings and concentrated work,
- Parking space or charging station, and
- Canteen/cafeteria.

This transparency becomes even more valuable the more comprehensive a company establishes a hybrid work model as an organizational form. The app defines a hybrid work model from a functional perspective as a work location arrangement for employees who do not have to spend their contractually specified working time entirely in company facilities (hereinafter referred to as *mobile work*).

By controlling hybrid work through an app, a company can efficiently plan the use of its facilities despite increased complexity. The person responsible for planning can not only assign workplaces and work zones to employees, but also block them so that, due to legal or operational distance regulations, upcoming renovation measures, etc., the use by the employees can be controlled. Fig. 2.17 shows, for example, delibarate blocking of workstations. Only the green workstations can be booked.

Fig. 2.16 Contacts/users and workplace bookings

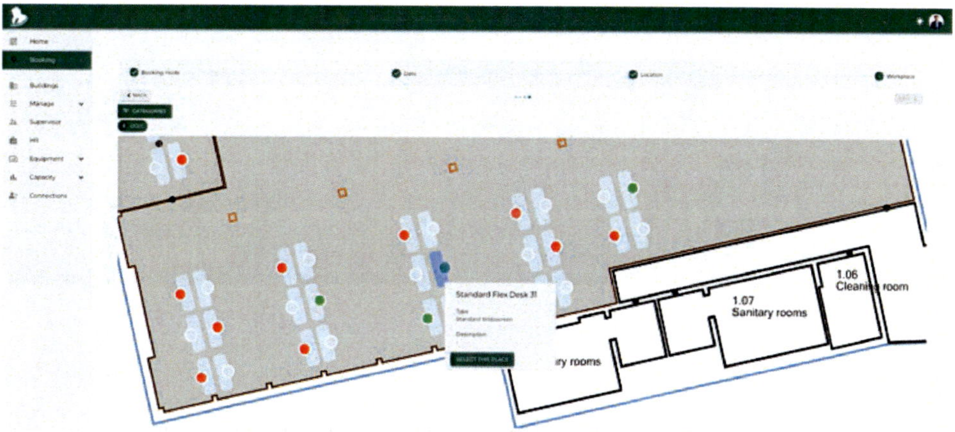

Fig. 2.17 Planning of complex workplace relationships

Furthermore, the company receives information about with whom the employees primarily work together, at which locations they can and want to work preferably or which type of workplaces, conference rooms or think tanks are booked intensively. Of course, this only applies if the employees release their usage data.

In addition to controlling and planning the use of office buildings, mobile work of employees can also be coordinated and subsequently analyzed, as far as this is legally allowed and organizationally possible. The planning and control of work inside and outside the office creates a complex optimization problem for workplace utilization, which the monitoring system solves and provides recommendations for better utilization in the future. As shown in Fig. 2.18, monitoring is required on different levels. This ranges from visualizing the utilization of individual workstations to highly aggregated location analyses.

The aim of workplace management systems like Seedit is to offer efficient digital solutions for this optimization problem. This ranges from the manual and fully automated booking possibility through intelligent mathematical algorithms that orient themselves to the preferences of the employees when choosing the workplace, to the creation of fully automated space utilization strategies. The latter can include aspects such as company-wide regulations for mobile work or the automatic adaptation of the equipment for bookable areas, so that the workplace offer adapts dynamically to the demand. Assigned spaces for departments can easily be checked. As a consequence of the algorithm, a department that only uses its four-person office 25% of the time can be moved to a two-person office. Thus, the app combines the analysis aspect of the complex optimization problem with the occupancy planning and creates a continuous optimization cycle, starting with the planning, followed by the organization of work on the spaces via book-

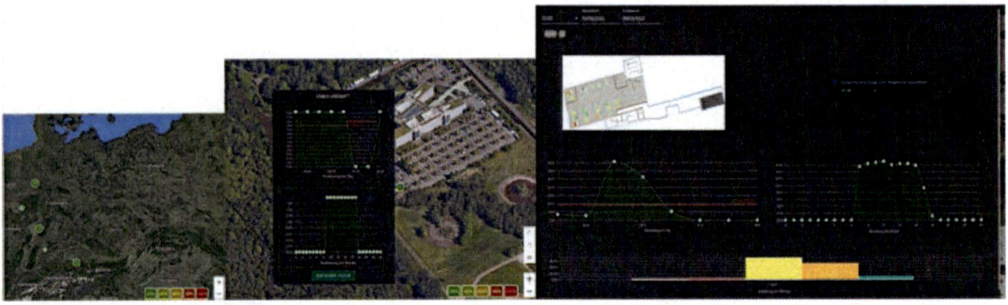

Fig. 2.18 Monitoring as the basis for optimizations

ings, up to monitoring and determining optimization possibilities in the space requirements analysis, which in turn influences the strategic space planning.

2.10.3 BIM for Digital Workplaces

It is to be examined whether and to what extent BIM can be used in connection with the use of digital workplaces in a meaningful way. Basically, this is conceivable in particular for the visualization of buildings, rooms and spaces in which digital workplaces are installed. Workplace relations, current occupancy rates, equipment characteristics and other attributes can play a role. The question of the sense of such applications can be answered just as well for new and converted buildings with existing BIM data as for existing buildings with traditional CAD data as well as for BIM in operation in general. However, in the case of digital workplaces, the monetarily measurable benefit of BIM will play an even greater role after a short time. This could also be the reason why many WMS do not use a detailed visualization (e.g. using CAD), but only a schematic representation with workplace symbols on a non-necessarily to-scale floor plan (e.g. in pdf or png format). As a result, there are hardly any examples of using BIM for digital workplaces, although this is at least possible with many IWMS in principle.

One of the few examples of the use of BIM in workplace management identified by the authors comes from Xavier University in the USA (see Haines and Norin 2016; May 2018a). The floor plan shown in Fig. 2.19 is derived from the BIM as-built model and contains the workplaces, the area labeling and dimensions.

The BIM-FM model was derived from the BIM as-built model. The following modifications were made during the creation of the BIM-FM model:

- Information that is irrelevant to FM has been removed, such as constructive details and execution plans. This information can be derived from the as-built model if required. This prevents the BIM-FM model from being overloaded with unnecessary information.

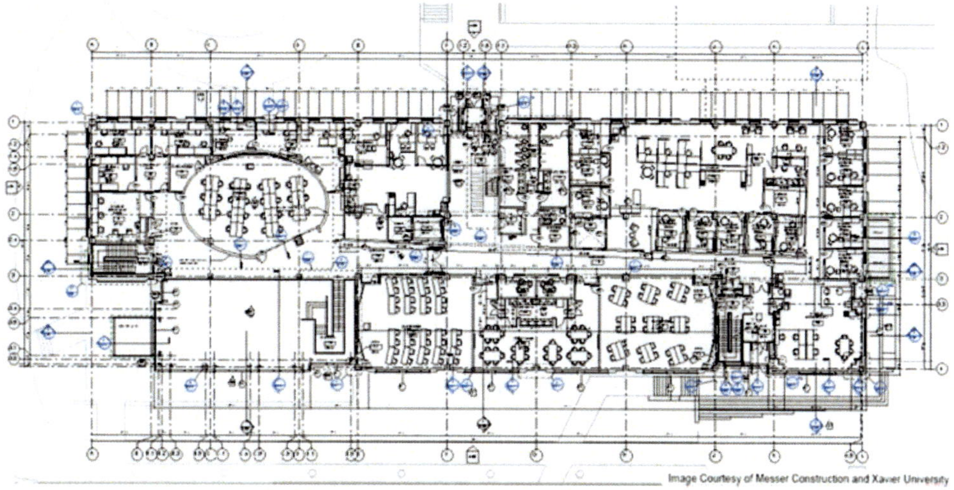

Fig. 2.19 Visualization of workplaces in a floor plan

- If linked models were used to explicitly represent the building structure, the building envelope or tenant installations, these were transferred to a single model.
- Where possible, linked models of the HVAC, for fire protection and use-specific building equipment (e.g. security) were also transferred to the FM model. However, for very large buildings, it may make sense to maintain several specialized models separately and continue to use existing links.
- Occupancy room numbers were taken over from the construction room numbers and the appropriate signage was designed directly (including electronic signs/monitors).
- For office space, workplaces and offices were separately numbered from rooms using an occupancy algorithm. This is necessary in order to assign workplaces to office users and to charge them (see Fig. 2.19).
- MEP elements were assigned special asset ID's (e.g. according to ASHRAE, for Germany according to DIN 276).
- The BIM-FM model was linked to a CAFM/IWMS which, for example, controls work orders, maintenance activities, floor space usage (see Fig. 2.20) and occupancy, repair and material costs.

2.10.4 Outlook

The rapid spread of autonomous and/or Integrated Workplace Management Systems (IWMS/CAFM) during the transformation of work, which was triggered by the worldwide COVID-19 pandemic, has set off a wave of digitalization in the use of real estate in many countries, which is further fueled by ESG (Environmental Social Governance) and the associated reporting. This development opens up new options for the use of BIM

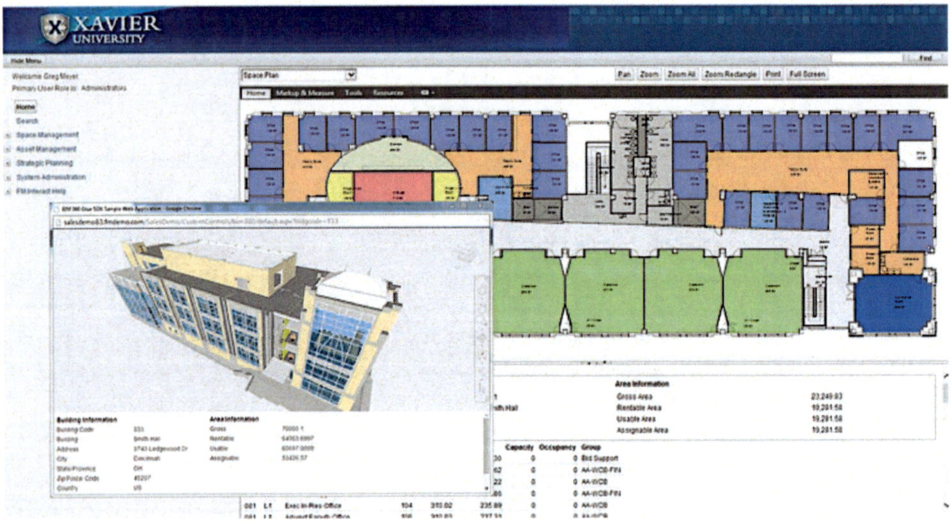

Fig. 2.20 3D model and 2D floor plan derived from the BIM-FM model with room numbers and space types

in FM. Whether this will become the case in most countries remains to be seen, especially in view of the lack of availability of BIM data for building operation, the existing time pressure to implement workplace management systems and the general reluctance to invest in the digitalization of FM in contrast to asset and property management (cf. NN 2021k). So it is not surprising that the authors could hardly find practical examples for the use of BIM in workplace management. Most examples come from North America and Asia.

2.11 Building Simulation

Whenever a real situation is too complex to describe the behavior of a building or a property (usually by means of mathematical-physical descriptions), simulation methods are used. A simulation always relies on a model (simulation model), which abstractly and as realistically as required for the respective simulation task maps real situations and scenarios. In contrast to simulations on a real system (e.g. crash tests in the automotive industry), only computer simulations are considered, in which the models are mapped to software and subsequently the simulation task is solved at least approximately by suitable mathematical methods and algorithms.

For this task, various simulation tools are used, which, however, all rely on some kind of modeling real building situations. It seems reasonable to consider these very different simulation models in the context of digital building models.

The aim of this section is to give an overview of the various fields of application in which simulation tools are used throughout the building lifecycle and to explain the interaction with digital building models. By linking with simulation models, the initially more static descriptive digital building models are extended by *What-if*-scenarios and thus made dynamic. In combination with the monitoring of measured values from building automation or IoT sensors (see Sect. 2.8), an actual digital twin of the building is only created in combining digital building models, simulation models and monitoring data (see Sect. 4.1).

2.11.1 Objectives of Simulation

In the lifecycle of a building, simulations are used in very different fields of application. The following fields of application of building simulations are further elaborated without claiming to be complete:

- Energetic simulation,
- Acoustic simulation,
- Lighting simulation,
- Flow simulation and building climate simulation,
- Structural simulation/structural planning,
- Construction process simulation,
- Operational simulation,
- Person flow simulation and
- Space simulation.

In general, all the simulation approaches have the following goals in common, of which only a selection is pursued in a single project:

- They *describe* the building or a section of the building in their field of application in a simulation model,
- They enable time-varying, dynamic *analyses* of the simulation model,
- They allow the *explanation* of a behavior of the building or of processes in a building and
- They enable the *prediction* (forecast) of future behavior of a building and its components and thus support decision making by humans.

Consequently, depending on the goal, the resulting simulation models are also differentiated as description, analysis, explanation and prediction models. Together, these different simulation models already allow for improvements to be made during planning, but also during later operations, by evaluating different variants (*what-if* scenarios) and thus reducing the risk of wrong decisions at an early stage of planning.

One of the major challenges for the reliable use of simulations is the calibration of the simulation methods used. By calibrating the simulation method, a simulation result is compared with reality in order to derive simulation parameters from it. These can be expected to produce a reliable simulation result in future applications of the simulation method. In this respect, the development of the simulation model and simulation method is usually an iterative process.

2.11.2 Integration of Simulation Tools with BIM

Digital building models that have been created using the BIM method already include numerous specialist information from different engineering disciplines, which are linked to the building components in the model. These are usually parametric BIM objects that are stored semantically in the model together with the relationships with each other. For example, the relationships between walls, rooms, floors and doors are already available in the model. For this reason, digital building models provide very good conditions for structuring and partially providing important information for simulation tools automatically.

2.11.2.1 Integrated Simulation Tools

Simple simulation tools, usually for a rough, approximate design, are already integrated into common BIM authoring tools (see Sect. 4.2). A good example of this are simple, usually static energy demand calculations, which work directly with the BIM models created in software tools such as Revit or ArchiCAD. This is usually achieved by extending the aforementioned BIM objects with simulation-relevant information attributes. The simulation tool thus integrated does not require its own model.

2.11.2.2 Simulation Tools with Independent Simulation Models

For comprehensive, more complex simulation tools, such as the building simulation system IDA ICE or software systems for structural design, which work, for example, using the finite element method (FEM), a specific, independent simulation model is required. Often, these simulation models contain more extensive, detailed information compared to the original digital building model, but sometimes the opposite is the case. For example, when simulating energy demand and user behavior, a high degree of geometric detail of the building components is not necessarily required, whereas very precise component and material properties are required.

The example of structural design clearly shows that transforming the original building model (e.g. the architectural model) into an analytical presentation is required for the simulation model. Fig. 2.21 (Trzechiak 2017) illustrates this by using the transformation of a digital capture of the as-built situation of a building by a 3D point cloud, which is transformed into an analytical model via a geometric architectural model. In this example, the geometric building elements of supports and beams are represented as 3D FEM

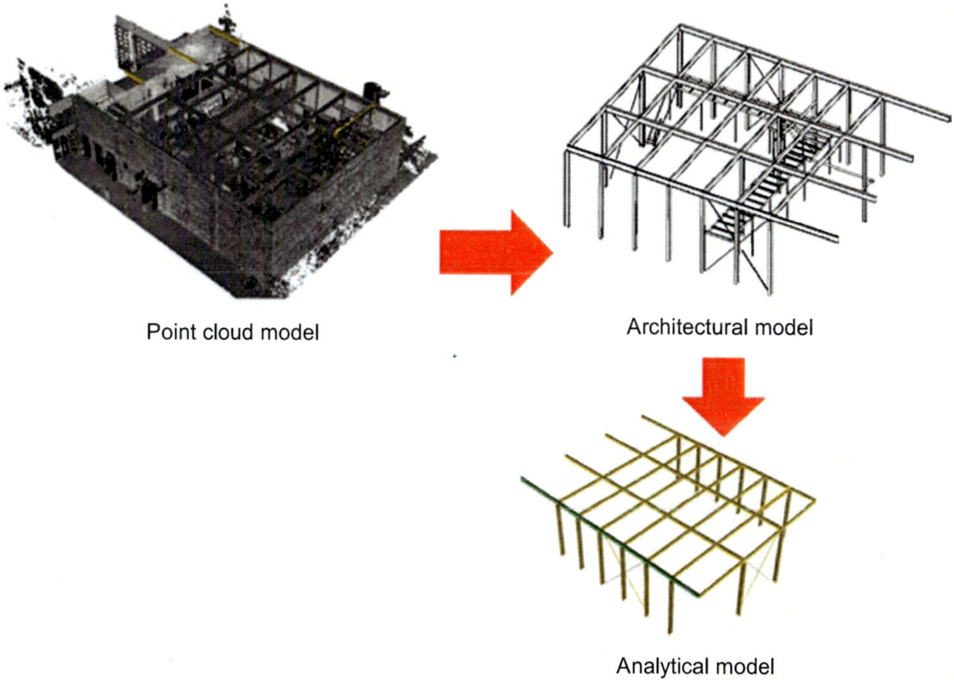

Fig. 2.21 Example of a model transformation for structural design

mesh (volume elements) or as shown in Fig. 2.21, as simple bar or beam elements, depending on the choice of calculation method. Often, manual rework is required in the simulation model as part of the model transformations, for example because connection points and transfer conditions were not sufficiently accurate in the original model.

The workflow of a BIM-based simulation is mostly designed as a control loop. Not only information from the digital building models is transferred to the simulation tool via interface, but also results of the simulation can lead to modifications of the original model. In this way the calculated energy demand of a room is written back to the room object in the original building model as attribute information. This also allows other disciplines to build on the simulation results and carry out their further specialist planning without directly accessing the simulation model. However, if the geometric dimensions of a component change because of the simulation, manual adjustments in the original model are usually required.

Basically, open data standards such as the IFC format are available for exchange with the simulation model. For individual simulation tasks, specific IFC specifications have already been defined for exchange (Model View Definitions—MVD). For exchange with simulation programs, the open Green Building eXtensible Markup Language schema (gbXML) has a special significance. In addition to geometric information, gbXML also exchanges other information important for energy simulations, such as usage profiles, lighting objects or weather data (cf. Sect. 5.3).

2.11.3 Applications

The following application fields for building simulations differentiate between the planning and operational phases. The use of building simulations in the planning phase are of particular interest from the perspective of real estate and FM for two reasons. First, building simulations that are used as early as possible in the planning cycle enable the choice of realistic simulation parameters to better consider the behavior of the building in later operations. In this way, the implementation of an FM-compliant planning that has been demanded for a long time but rarely practiced becomes more realistic. Secondly, all application fields mentioned below for the planning phase can also occur in the operation phase for renovation, refurbishment or conversion measures.

2.11.3.1 Building Simulation in the Planning Phase

Depending on the planning discipline, different areas of building simulation play a role, which are often based on similar discrete or continuous simulation approaches and methods. Typical simulation tools such as TRNSSYS, IDA ICE, EnergyPlus,Top Energy or Greenbuilding Studio usually support several areas of application and can, for example, support simulations of building behaviour with regard to electrical or thermal energy profiles. In some cases, the tools also master simulation methods for acoustic behavior, building or room climate or lighting scenarios.

The application of simulation tools in the field of energy demand determination serves basically to predict the primary energy demand, as required, e.g. by legal requirements in the Energy Saving Ordinance (NN 2013a). Other simulation purposes are the determination of heating and cooling loads for the design of HVAC systems and thus for ensuring a comfortable building climate, without unnecessarily over-dimensioning the systems and thus taking unnecessary higher energy consumption into account.

Dynamic simulations contribute significantly to determining more precise load profiles with realistic assumptions of user behavior, to reduce the often occurring difference between the calculated energy demand and the energy consumption measured subsequently. In the field of lighting analyses, building simulations can be used to design shading and light-directing systems which, through improved daylight utilization further reduce the primary energy demand of the building (cf. also Kolk 2021). Against the background of open-plan office concepts, such as open space approaches (cf. also Sect. 2.10), the simulation of the acoustic characteristics of rooms and the prediction and optimization of the air flow behavior of ventilation and air conditioning have gained great importance to guarantee draught-free workstation environments here.

For further information, also on the previously used example of structural design or structural simulation, reference is made to the further explanations of Fink (2015) and Treeck et al. (2016).

Dynamic simulations that map the technical control behavior of technical systems are increasingly considering the interaction with energy supply grids. The interaction of the energy-efficient building systems with the energy grids is of central importance against the background of the ever-increasing share of renewable energies in the future. Due to the dependence of the production of renewable energy from wind and solar energy on weather conditions, a grid-compatible building operation will be required to ensure grid stability. Taking into account sector coupling the operation of buildings will have to adapt to energy demand fluctuations in energy networks. In this context, corresponding storage/buffer systems will play a central role. Voss et al. (2021) describe an interesting simulation approach in which the thermal storage capacity of buildings is used to make the building itself usable as a storage for the fluctuating energy production of regenerative energy sources. For example, if there is an excess of regenerative energy, the building is only slightly warmer, without affecting the comfort of the building users. If there is a shortage, the energy stored in the building is taken out again, in the simplest case by reducing the amount of energy used for heating. The authors show that with easily measurable KPI's (e.g. the temperature on ceiling surfaces) the storage state of the building can be calculated and thus a grid-compatible operation of the building is possible with this favorable storage technology.

2.11.3.2 Construction Process and Operation Simulation

Another area of application for dynamic simulations are construction process simulations. These simulations, also referred to as 4D simulations, use the ability of BIM models to automatically perform mass or quantity takeoffs based on the BIM objects contained therein. If these component-oriented analyses are connected with the work packages of a construction schedule (project planning), in which the components are procured or assembled on the construction site, for example, the geometric dimensions (3D) of the BIM model extended by the time dimension (4D). In this way, the BIM model "grows" according to the activities in the project, with more and more BIM components being added in the corresponding construction phases. If, in parallel to this sequence of events, the cost curves (or payment flows) are also considered, a 5D simulation is the result (cf. Fig. 2.22, Gruschke and Werner 2013).

Such construction process simulations enable for example, to avoid schedule conflicts, optimize construction site logistics, but also to facilitate the acceptance of construction services during construction by comparing the planned with the actual progress of construction. The same applies in the context of commissioning planning (cf. Sect. 3.3).

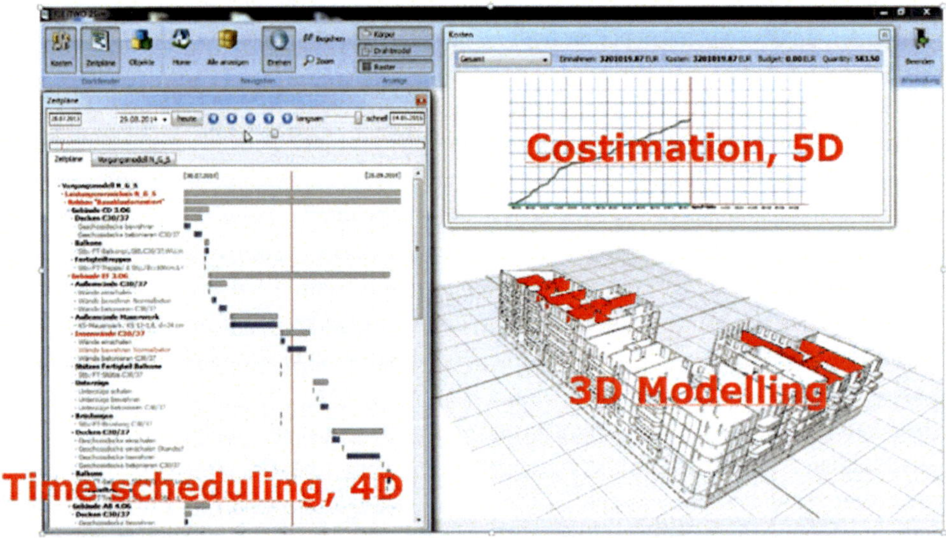

Fig. 2.22 Example of a 5D construction process simulation

The field of application of 4D and 5D simulations can be extended by further dimensions in coupling the BIM model with real-time sensor data from the building, for example by means of IoT devices or by a building management system resulting in a digital twin (cf. Sect. 4.1). Hence, operational simulations are enabled.

2.11.3.3 Space and People Flow Simulation

For many organizations, space costs are the second-largest cost block after personnel costs. Spaces have a sustainable impact on the behavior of people and the environment, but are often not used economically. The responsible use of this resource is a social concern. Consequently, the optimization of space resources is increasingly supported by space simulations(cf. May 2016a). This awareness has gained considerable importance during the Covid-19 pandemic.

Because BIM models usually already contain categorized spaces and rooms, they are a very good starting point for formalized space simulations. In this way, the spatial proximity, accessibility or the visibility of places (e.g. room or room zone) from a certain location can be investigated in the space simulation. For this purpose, spatial neighborhood relationships (adjacencies) and movement possibilities are taken into account in the form of graphs in the simulation model. Sometimes simplified raster models (grids) are also used, which represent a special case of graph models. Connections between rooms are then mapped as edges of the graphs. This modeling is often refined by adding additional nodes such as passages, doors, stairs or elevator shafts. Algorithms can now be applied to these graphs, for example to determine the shortest, fastest or most economic route.

Based on these simulation models, various simulation tools have been developed for space optimization and people flow simulation. With the Recotech method (cf. May 2016b; NN 2021al), the room occupancy can be optimized automatically according to different criteria such as communication costs and distances.

The Space Syntax method (cf. May et al. 2013; NN 2021am) is used to simulate movement patterns and, for example allows to predict the behavior of visitors in a public building and thus, to support the development of guidance (navigation) systems. In recent years, 3D simulations have been used more and more, in which the spatial models are derived from BIM and are partly analyzed using expert systems (cf. Li et al. 2009; Cho and Kwon 2021).

Further agent-based people flow or crowd simulation systems like crowd:it (cf. NN 2021an) are also used for testing usage concepts, developing evacuation scenarios for fire protection concepts, or directing the flow of people to ensure hygiene and pandemic concepts.

2.12 PropTechs

PropTechs refer to companies that support and drive the digital transformation of the real estate industry (usually using innovative hardware and/or software). This means not only digitizing processes, but also using new methods and technologies to optimize existing or new processes, provide new products or services, and even replace outdated business models.

The term PropTech is a combination of the terms property services (real estate services) and technology and can be translated as real estate technology or technology for the real estate industry. *Technology* is always used in this context in the sense of information technology or digitalization.

In other industries, similar terms have been established with the buzzword "Tech", such as FinTech in the financial sector or InsurTech for insurance.

Most PropTechs use technologies that have been known in computer science for many years (e.g. AI, lexical analysis, pattern recognition), but are only now practically available and affordable. In this respect, the innovation risk in the PropTech scene is not as high as, for example, in the field of biotechnology.

PropTechs often specialize in narrowly focused topics and offer corresponding services or products with innovations, including data management, real estate management, visualization and planning, including BIM technologies. Fig. 2.23 shows the distribution of PropTechs in the lifecycle of the property according to the PropTech Germany study (NN 2021o).

In Germany there is a particularly diverse PropTech scene, which is surprisingly well organized compared to other HighTech hotspots (USA, Far East). This may also be due to the fact that it is relatively difficult for startup companies in Germany to obtain ven-

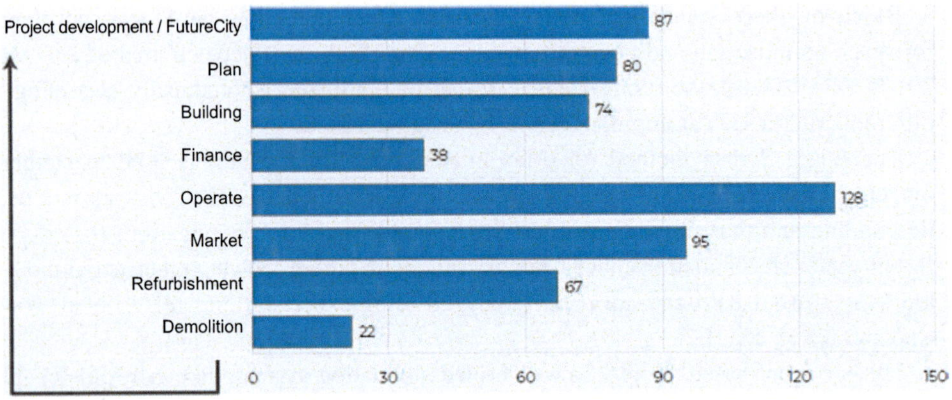

Fig. 2.23 Distribution of PropTechs on real estate tasks

ture capital for the first financing rounds after the company is founded. Therefore, the PropTech networks offer support for self-help.

It is interesting that in Germany most PropTechs, as the term PropTech suggests, are active in the various fields of property management and hardly operate in FM. This also applies to most companies mentioned in the PropTech market overview (NN 2021l) under the heading "Real Estate Management". The almost complete abstinence of start-ups in the field of FMTech (in USA: FMIT) could be due to the fact that in the German FM market only minimal profit margins are achieved and even after the end of the COVID-19 pandemic there are hardly any higher investments in digitalization initiatives. Fig. 2.24 (NN 2021k) from the GEFMA/LÜNENDONK CAFM-Trendreport 2021

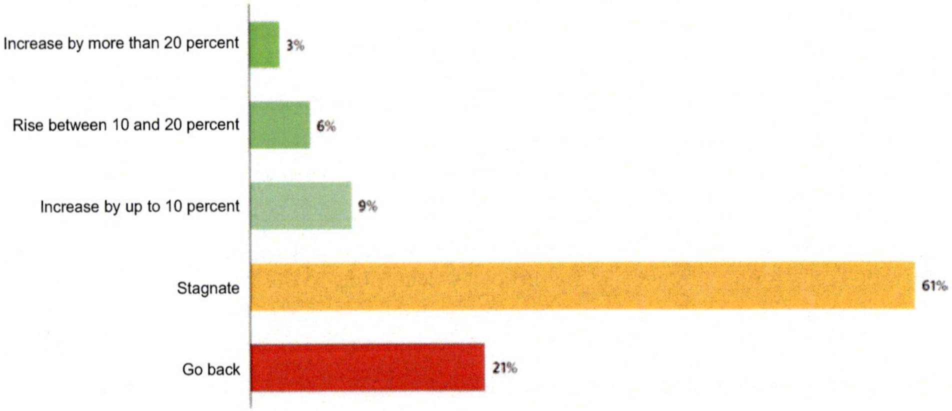

Fig. 2.24 Impact of the COVID-19 pandemic on budget planning in 2021/22

shows the summary of the responses of over 100 companies to the question of the development of FMIT budgets in the years 2021/2022.

The PropTech scene is an important factor in the digital transformation of the real estate industry. Through it, innovations are created and implemented in practical solutions.

2.13 Summary

Digitalization is an important, in many cases even the most important, driver of innovation in almost all areas of industry. This also applies to the real estate industry and, in particular, to real estate and FM.

In this context, established technologies such as CAFM work closely with modern digitalization trends such as IoT. In cases where it is possible to integrate different IT-based technologies, the economic benefit can be achieved easily, because costly and error-prone transformation and coordination processes fall away.

In this chapter, digitalization trends were introduced which are already important for the real estate industry today, but even more in the future. This should not only illustrate the variety of technologies and their development potentials to experts, but primarily to interested people using the technology in practice.

Technologies and systems such as CAFM/IWMS have developed over the past three decades into proven and indispensable tools and platforms that have their strength in the use phase of buildings. These are now supplemented by a variety of BIM tools and platforms, which helps to close the information gap between planning/construction and use/operations. The shift of many mentioned applications to the cloud is a trend that has been observed in other industries for some time and is now also reaching the real estate sector. Buildings and their technical systems produce large amounts of state data (big data). If these are captured by suitable sensor systems and measuring devices and are stored digitally, completely new evaluation options (big data analytics) arise, which, for example, allow AI/ML-based predictions to be made about the future behavior of building systems or their users.

Even before buildings and their systems are erected and put into operation, statements can already be made about their behavior. For this purpose, there are a variety of simulation techniques available, which are generally based on mathematical-physical models and the associated solution methods. The use of such methods such as thermal-energetic, space or person flow simulations is already common practice in the planning phase. In operation, there are first approaches such as occupancy and move simulations, which also have an impact on the digital workplace. However, there is still a high need for catch-up here, which is associated with considerable research effort but also great prospects for success.

The list of digitalization technologies is certainly longer than presented in this chapter. Technologies such as blockchain, robotics or even autonomous driving obviously also have an impact on the real estate industry, be it on contract and service management or even logistical tasks. And even technologies such as Quantum Computing (see Heßling 2017), whose use apparently lies in the distant future, will not bypass real estate-related (optimization) tasks as soon as it is recognized that selected, extremely complex tasks can be solved exactly in fractions of the time of previous approaches or become tractable at all.

It is to be expected that the fascination and efficiency of the use of digitalization technologies and the pursuit of new trends will lead to further modernization and higher efficiency of real estate and facility management.

References

Altmannshofer R (2018) Künstliche Intelligenz im FM. Der Facility Manager 25(2018)1/2, 50–51

Bollmann T, Zeppenfeld K (2015) Mobile Computing—Hardware, Software, Kommunikation, Sicherheit, Programmierung. 2. edn. 2015, W3L AG, Dortmund, 216 S

Borrmann A, König M, Koch C, Beetz J (Eds.) (2018) Building Information Modeling—Technology Foundations and Industry Practice. Springer Nature, 2018

Buxmann P, Schmidt H (Eds.) (2019) Künstliche Intelligenz—Mit Algorithmen zum wirtschaftlichen Erfolg. Springer Gabler, 2019, 206 p

Cho J, Kwon O (2021) BIM Space Layout Optimization by Space Syntax and Expert System. Korean J. of Computational Design and Engineering 22(2017)1, 18–27

Ellmer D, Salzmann P (2014) Augmented Reality im FM. Facility Management 2014/15—Das Branchenjahrbuch. F.A.Z-Institut, Frankfurt, 2014, 84–95

Fink T (2015) BIM für die Tragwerksplanung. In: Borrmann A, König M, Koch C, Beetz J (Eds.). Building Information Modeling. Technologische Grundlagen und industrielle Praxis. Springer Vieweg

Göring M (2017) Begegnung mit einer unbekannten Art. National Geographic, Juli 2017, 58–81

Gruschke M, Werner P (2013) Intelligente Planung und Kostenkalkulation am virtuellen Gebäudemodell unter Anwendung von Building Information Modeling (BIM), Master Thesis, HTW Berlin

Haines B, Norin R (2016) Utilizing distributed BIM based Workplace Management tools to analyze spatial performance of an entire facilities portfolio. Autodesk University, 37 p

Hanhart D (2008) Mobile Computing und RFID im Facility Management—Anwendungen, Nutzen und serviceorientierter Architekturvorschlag. Springer-Verlag Berlin Heidelberg, 213 p

Heßling H (2017) Quantum Computing—A Digitization Option for FM? Tagungsband INservFM, Frankfurt, 21.–23.02.2017, p 613–625

Hoar C, Atkin B, King K (2017) Artificial intelligence: What it means for the built environment. RICS Report, October 2017, 28 p

Hofstadter DR (1985) Gödel, Escher, Bach—ein Endloses Geflochtenes Band. 5. edn., Klett-Cotta, 1985, 844 p

Hwang K (2017) Cloud Computing for Machine Learning and Cognitive Applications. The MIT Press, Cambridge, London, 2017

Kolk D (2021) Der BIM-BOOM geht weiter—Für Leuchtenhersteller und Lichtplaner lohnt sich BIM mehr denn je. Licht (2021)8, 60–61

Li Y, Lertlakkhanakul J, Lee S, Choi J (2009) Design with Space Syntax Analysis Based on Building Information Model: Towards an interactive Application of Building Information Model in early Design Process. In CAADFutures, Les Presses de l'Université de Montréal, Montreal, QC, Canada, 502–514

Lösel S (2017) Was ist Mobile Computing? (01.08.2017), https://www.it-business.de/was-ist-mobile-computing-a-634341/ (retrieved: 23.08.2021)

May M (2016a) Flächeneffizienz durch Analyse, Simulation und Optimierung. In: Knaut M (Ed.) Digitalisierung: Menschen zählen—Beiträge und Positionen der HTW, BWV Berliner Wissenschafts-Verlag, 282–287

May M (2016b) Best Practice Space Optimisation for Office Buildings. Corporate Real Estate Journal 5(2016)2, 154–170

May M (2017) BIM-based Augmented Reality for FM. FMJ (USA), 27(March/April 2017)2, 16–21

May M (Ed.) (2018a) CAFM-Handbuch—Digitalisierung im Facility Management erfolgreich einsetzen. 4th edn., Springer Vieweg, Wiesbaden, 2018, 713 p

May M (2018b) Artificial Intelligence and Machine Learning in FM. eFMinsight (June 2018)45, 8–10

May M (2020) Generatives Flächendesign. Der Facility Manager 27(April 2020)4, 28–33

May M, Williams, G (Eds.) (2017) The Facility Manager's Guide to Information Technology—An International Collaboration. 2nd edition, IFMA, Houston, 2017, 635 p

May M, Kohlert C, Schwander C (2013) Raumforschung mit Space Syntax—Neues (CA)FM-Geschäftsfeld. Der Facility Manager 20(Januar/Februar 2013)1/2, 48–52

May M, Clauss M, Salzmann P (2017) A Glimpse into the Future of Facility and Maintenance Management: A Case Study of Augmented Reality. Corporate Real Estate Journal 6(2017)3, 227–244

Mell P, Grance T (2011) The NIST Definition of Cloud Computing. National Institute of Standards and Technology, Gaithersburg, September 2011, Special Publication 800-145, http://nvlpubs.nist.gov/nistpubs/Legacy/SP/nistspecialpublication800-145.pdf, (retrieved: 18.06.2021)

Milgram P, Takemura H, Utsumi A, Kishino F (1994) Augmented Reality: A class of displays on the reality-virtuality continuum. SPIE Proceedings Vol. 2351: Telemanipulator and Telepresence Technologies, Boston, 1994, 282–292

NN (2009) Cloud Computing – Evolution in der Technik, Revolution im Business. BITKOM-Leitfaden, Oktober 2009. https://www.bitkom.org/Publikationen/2009/Leitfaden/Leitfaden-Cloud-Computing/090921-BITKOM-Leitfaden-CloudComputing-Web.pdf (retrieved: 18.06.2021)

NN (2013a) EnEV—Energieeinsparverordnung. https://www.bmwi.de/Redaktion/DE/Downloads/Gesetz/zweite-verordnung-zur%20aenderung-der-energieeinsparverordnung.html (retrieved: 30.10.2021)

NN (2014a) GEFMA Richtlinie 410: Schnittstellen zur IT-Integration von CAFM-Software, Juli 2014, 11 p

NN (2016b) Cloud Computing im Facility Management. White Paper GEFMA 942, 01.11.2016

NN (2021a) GEFMA Richtlinie 400: Computer Aided Facility Management CAFM—Begriffsbestimmungen, Leistungsmerkmale, März 2021, 19 p

NN (2021g) https://de.wikipedia.org/wiki/Internet_der_Dinge (retrieved: 27.05.2021)

NN (2021h) https://internetofthingsagenda.techtarget.com/definition/Internet-of-Things-IoT (retrieved: 27.05.2021)

NN (2021i) https://www.bsi.bund.de/DE/Themen/Unternehmen-und-Organisationen/Informationen-und-Empfehlungen/Empfehlungen-nach-Angriffszielen/Cloud-Computing/Grundlagen/grundlagen_node.html (retrieved: 18.06.2021)

NN (2021k) CAFM-Trendreport 2021—GEFMA 945, GEFMA/LÜNENDONK, Juni 2021, 63 p

NN (2021l) https://proptech.de/wp-content/uploads/2021/04/PropTech_Uebersicht_Maerz_2021.pdf (retrieved: 27.06.2021)

NN (2021o) PropTech Germany 2021 Studie. https://proptechgermanystudie.de/ (retrieved: 27.06.2021)

NN (2021aj) PropTech-Unternehmen. https://proptech.de/ (retrieved: 24.10.2021)

NN (2021ak) Seedit. https://recotech.de/overview/seedit/ (retrieved: 24.10.2021)

NN (2021al) Recotech-Flächenoptimierung, https://recotech.de/overview/recotech/ (retrieved: 01.11.2021)

NN (2021am) Space Syntax. https://www.spacesyntax.net (retrieved: 01.11.2021)

NN (2021an) cowd:it-Personenstromsimulation. https://www.accu-rate.de/de/software-crowd-it-de/ (retrieved: 01.11.2021)

NN (2022a) IoT im Facility Management, White Paper GEFMA 928, 2022

NN (2022c) GEFMA Richtlinie 410: Schnittstellen zur IT-Integration von CAFM-Software, Februar 2022, 12 p

Schneider U (Ed.) (2012) Taschenbuch der Informatik. 7. Auflage, Carl Hanser Verlag München, 2012

Teicholz P (Ed.) (2013) BIM for Facility Managers. John Wiley & Sons, Inc., Hoboken, New Jersey, 2013

von Treeck C, Elixmann R, Rudat K, Hiller S, Herkel S, Berger M (2016) Gebäude. Technik. Digital. Building Information Modeling. Springer Vieweg, 453 p

Trzechiak M (2017) The BIM-to-FEM Interface—Development of Computational Tools for BIM Data Exchange and Check. Master Thesis, HTW-Berlin

Vaughan G (2020) Event-based Microservices: Message Bus—Simple, Scalable, and Robust. https://medium.com/usertesting-engineering/event-based-microservices-message-bus-5b4157d5a35d (retrieved: 26.11.2021)

Voss M, Heinekamp J, Krutzsch S, Sick F, Albayrak S, Strunz K (2021) Generalized Additive Modeling of Building Inertia Thermal Energy Storage for Integration Into Smart Grid Control. IEEE Access 99(May 2021)1, 71699–71711

BIM Basics for Real Estate and Facility Managers

3

Markus Krämer, Thomas Bender, Joachim Hohmann, Erik Jaspers, Thomas Kalweit, Michael Marchionini, Michael May and Matthias Mosig

3.1 From CAD to BIM

BIM is often referred to as the next level of Computer Aided Design (CAD) or Computer Aided Architectural Design (CAAD) and in the past was sometimes used synonymously with 3D-CAD. The following section aims to dispel this misunderstanding by highlighting important steps in the development of CAD systems and their methodological approach in order to clearly distinguish them from the current understanding of BIM (cf. Fig. 3.1).

The basic idea of the first CAD systems was to be able to create paper drawings more efficiently using a computer. In this sense, CAD was referred to *Computer Aided Drafting*. Nonetheless, CAD software already used internal data models for this

M. Krämer (✉)
Hochschule für Technik und Wirtschaft Berlin, Berlin, Germany
e-mail: markus.kraemer@htw-berlin.de

T. Bender
pit – cup GmbH, Heidelberg, Germany
e-mail: thomas.bender@pit.de

J. Hohmann
Technische Universität Kaiserslautern, Kaiserslautern, Germany
e-mail: joachim.hohmann@bauing.uni-kl.de

E. Jaspers
Planon B.V., Nijmegen, Germany
e-mail: erik.jaspers@planonsoftware.com

T. Kalweit
net-haus GmbH, Berlin, Germany
e-mail: t.kalweit@net-haus.com

M. May et al. (eds.), *BIM in Real Estate Operations*,
https://doi.org/10.1007/978-3-658-40830-5_3

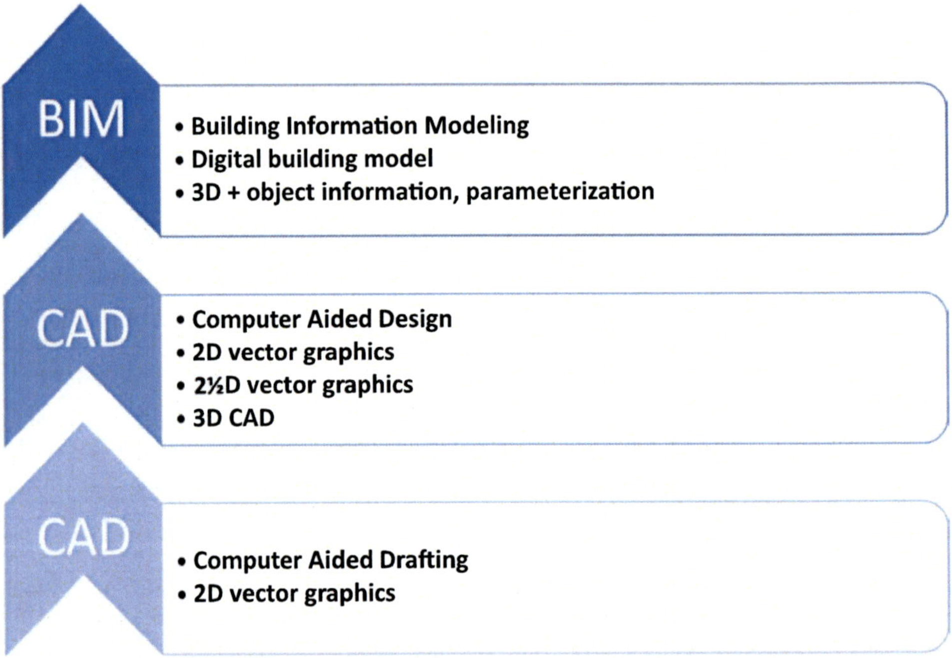

Fig. 3.1 Development from CAD to BIM

purpose. However, at first only simple 2D vector graphics were used, utilizing basic geometric elements (2D primitives) such as lines, arcs, circles and text. The major goal of this early CAD software was to automate the creation of (paper) drawings, which could then be plotted out. Additional attributes beyond geometric information can be expressed as line types, line widths and colors or layers, in accordance with established conventions of technical drawings, and stored in the CAD model.

M. Marchionini
ReCoTech GmbH, Berlin, Germany
e-mail: marchionini@recotech.de

M. May
Deutscher Verband für Facility Management (GEFMA), Bonn, Germany
e-mail: michael.may@gefma.de

M. Mosig
TÜV SÜD Advimo GmbH, Munich, Germany
e-mail: matthias.mosig@tuvsud.com

Even at early stages, it became obvious that efficiency goals (compared to drawing on a drawing board) could only be achieved by adapting the working methods. Consequently, later CAD systems offer added values (beyound drafting) by increasingly supporting design and construction activities. Drawing programs thus became design systems supporting users with libraries of standard symbols or parts and simple functions defining part dimensions or layout. The still common term *Computer Aided Design* takes this development into account. Only then, and with the availability of affordable hardware and software was CAD able to gain widespread acceptance. Today, classical drawing boards have almost completely disappeared from the professional life of architects and engineers. With this development, the CAD model became the key mechanism for information exchange between the parties involved, whereby the rules of technical (paper) drawings were still required for the interpretation of the models.

Further developments of modern CAD systems were characterized by increasingly complex CAD models that covered more and more aspects of the building. With regard to the geometric representation, basic 2D elements were first arranged in space, provided with a height indication (2½D) and finally extended by 3D surface and volume elements. Today's 3D CAD systems no longer represent the building geometry as a wireframe, but as a boundary representation or full-fledged solids. In fairly simple solid modeling (Constructive Solid Geometry), the building geometry is constructed by combining basic elements such as cubes, cylinders and spheres with standard logic operations such as union, intersection or substraction. For example, a door opening in a wall is formed by subtracting a cuboid (opening) from another cuboid (wall).

The next step in development prepared the today's understanding of digital building models in the sense of BIM. In a component- or object-oriented approach, existing building components are mapped to virtual objects in the CAD system. In addition to the 3D geometry of the component, they have a content-related (semantic) meaning, for example, belong to the class of doors, walls, windows or ceilings. Each component (object) thus comprises additional fact information, which is mapped as attributes.

For a digital BIM-based building model, such an object- and component-related internal representation is required. In addition, the creation of buildig parts is often controlled by parameters (parametric modeling) that interlink these building parts in a meaningful way. So, the building part "door" also stores the reference to the building part "wall", in which this door is positioned. If such a parametric model is moved, the corresponding door opening in the wall is automatically moved.

3.2 CAFM Basics

The development of Computer Aided Facility Management (CAFM) became important in the beginning of the 1990s. CAFM has become an established, reliable method and technology within this period when it comes to data management and the efficient control of processes in real estate and facility management (May 2018a).

In Germany, the topic of "IT and digitalization in facility management" is driven forward both scientifically and practice-oriented by the Working Group Digitalization of GEFMA (May 2021). In addition, a group of CAFM software providers, FM consultants and FM service providers, the CAFM RING e. V., are developing approaches to improve interoperability between CAFM products and BIM.

With its lifecycle-oriented view, CAFM is very closely related to Building Information Modeling (BIM). However, problems of system integration and interoperability still present challenges when planning the implementation of the BIM approach. A key problem was that there has never been a data exchange format with which one could migrate a complete CAFM solution with little effort from one system to another, although this issue is quite relevant in practice. Further progress is to be expected through the development of standardized exchange formats and integration technologies.

CAFM systems are powerful information tools for mapping, evaluating and controlling FM organizations and processes, as well as for documenting compliance-relevant processes.

In Sect. 2.1 the difference between CAFM software and CAFM system was already briefly discussed. Put simply: CAFM software is what you purchase from a CAFM software provider, while a CAFM system is what is developed in a CAFM project by the user organization—a ready-to-use and organization-specific IT system based on their own data and processes.

Each CAFM software has an internal structure that is aligned with the tasks of FM. This includes a FM-specific data model as well as a mapping of process flows and, in addition, various calculation algorithms and a reporting tool. CAFM software offers various basic functions that—depending on the task and vendor—are more or less sophisticated and allow the user to extend and customize the software to become a CAFM system.

The performance and functionality of CAFM software is constantly growing and meanwhile very impressive. The common supported application areas (CAFM core applications) have already been mentioned in Sect. 2.1. To ensure this, demanding requirements are placed on modern CAFM software systems, including:

- Extensive functionality,
- Support of FM processes (modeling and control),
- Openness and flexibility (e.g. with regard to data model, processes and reporting)
- Interfaces (I/O alphanumeric data, graphics, database, API, webservice, …),
- Multi-client,
- Data security,
- Ability to integrate into the company's IT environment,
- Accesibility via WWW,
- Mobile usage (support for mobile devices online and offline),
- Modern n-tier software architecture,
- Support of standardized data exchange formats,

- Extensive graphical and alphanumeric reporting options,
- Lowest possible costs of IT operations,
- Efficiency (handling of large amounts of data),
- Version and change history management,
- Ease of use (uniform and intuitive interface, responsive design),
- Easy learnability (also for non-IT experts) and
- High quality of the help system, assistants and accompanying documentation.

Figure 3.2 shows the most important requirements for a CAFM software with regard to application focus, the technology to be supported and the required flexibility to allow adaptations to be made by either the software provider or the user.

In addition to the requirements for functionality and usability, the system costs, the services offered, the user support as well as the vision and innovative strength of the software provider are important for the selection of a software.

As in the entire IT world, numerous new technologies are used in the development of CAFM software. In addition to proven architectures, the trend towards web-, cloud-based and mobile system usage is increasingly important. Occasionally you can still find

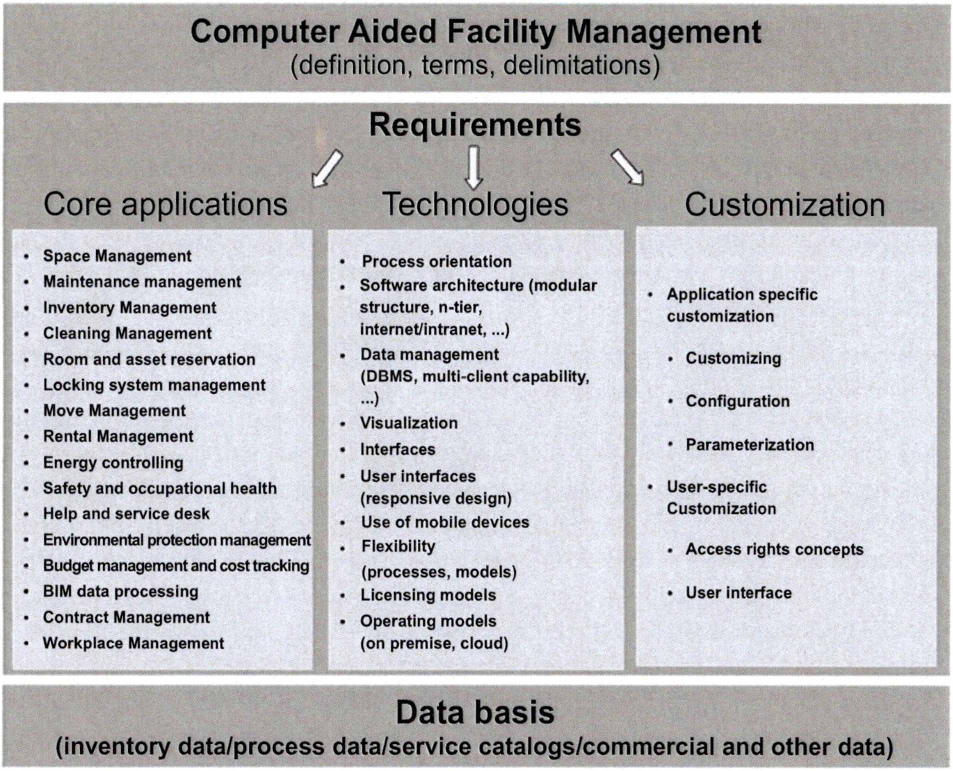

Fig. 3.2 Requirements for CAFM according to GEFMA guideline 400

pure client-server solutions. However, most of the CAFM software providers have recognized the benefits of multi-layered (n-tier) software architectures in terms of software development and maintenance and migrated their system architectures to modern concepts. This allows a separate implemenation of the data storage (data layer), the program logic (application layer) and the user interface (presentation layer) independently of each other. The variety of structures and uses of CAFM software is increasing. This leads to more complex systems with correspondingly growing demands on installation, operation and maintenance.

The correct procedure for introducing and using IT tools to support FM is an important success factor in CAFM projects (NN 2017a). Often it is unclear how to approach a successful CAFM implementation and how to prove or estimate the viability. There are now methods available that allow, for example, to identify and prioritize the value drivers of a CAFM implementation or to estimate the return on investment (ROI) or other economic key figures of a CAFM solution. A period of typically two to three years for the amortization of investments in CAFM is quite a key value in the application software industry. It is of interest to investigate to what extent the CAFM economic model can be transferred and adapted to BIM (cf. Chap. 6).

For operation and maintenance of a CAFM software, different approaches are available. For the selection, the number and spatial distribution of building users, internal capacities for data management and system administration/-operation, reliability requirements of system operations and the desired financing model must be considered.

In the past, CAFM software was almost exclusively operated by the user organization (on premise). They often provided the required hardware and network infrastructure, installed operating systems, databases and the CAFM software itself. In this traditional approach, the vendor of the CAFM software suplies the user organization with the CAFM software licenses, provides update services as well as hotline services for solving technical issues via a maintenance contract. An IT service provider for the CAFM software, often the software vendor itself, supports user organizations with consultancy, software customizing and training. Furthermore, services for data capture and if necessary, data preparation or migration is offered.

Internet technologies have encouraged many software providers to offer the hosting of complete CAFM solutions accessible over the Internet. Often, such solutions rely entirly on the use of web-based user interfaces. These different approaches, reflected in new ways of operation and maintenance, are summarized under the technical term Cloud Computing (Kalweit and May 2017, see also Sect. 2.5). Whether and to what extent a cloud solution makes sense depends on the security needs, requested flexibility, availability of human resources and their expertise. Almost all CAFM software vendors are offering those cloud solutions, where Software as a Service (SaaS) is the predominating business model. However, only a few software providers are currently using cloud solutions exclusively.

A common understanding of which functionality of a CAFM system is mandatory or preferrable is still missing and depends on the specific use case. Reasons for this include a lack of market transparency, the large variety of offers and, in some cases, a lack of knowledge.

The number of CAFM software systems offered in the German-speaking market has increased to over 60, offering the most diverse variety compared with the international market. More than 30 of these are presented in detail in the CAFM Market Overview (NN 2021b and NN 2022b), which is published annually. The market overview is the best choice for those interested in getting an overview of performance and focus of various systems. It is often used to pre-select the CAFM vendor under consideration. In addition, it also indicates which of the systems have received the CAFM certificate according to GEFMA guideline 444.

It is also interesting to compare certain aspects over a period of observation. Fig. 3.3 (cf. NN 2022b) shows an overview of the percentage of costs in a typical CAFM project. Particularly noteworthy is the decreasing amount of data acquisition costs due to new or improved capturing methods (cf. Sect. 5.2).

Banks and insurance companies as well as large and medium-sized manufacturing companies are among the pioneers in the use of CAFM. Increasing demand is shown, for example, in the public sector, retail, hospitals, utility companies, housing associations, airports but also among medium-sized companies.

Numerous successful CAFM projects in different industries are described in detail in May (2018a). Although the training situation in CAFM has improved in recent years, the CAFM expert who understands both the FM processes and basic digitalization skills is still rare and is urgently demanded by user organizations and service providers. GEFMA has revised the competence requirements for future facility managers in GEFMA guideline 610 (NN 2021c), with digitalization playing a very decisive role.

Some CAFM projects still fail today, which is often due to insufficient human resources and underestimated efforts for data acquisition, but rarely due to the availability of technologies and software systems.

As a successful contribution to quality assurance and to improve market transparency, the CAFM certificate according to GEFMA guideline 444 has been proven itself in practice (NN 2020a, see also Sect. 8.2), which was first published by the Working Group on Digitalization of GEFMA in 2010. This certificate attests to the tested software products the fulfillment of general technical and specific professional functions, which are currently documented in 17 test catalogs. On the GEFMA homepage, more than 20 certified or re-certified CAFM software products of domestic (DACH region) and other international vendors are listed. The certificate allows potential CAFM users to preselect software products that are of interest. The providers thus do not have to prove each basic function of their software to their potential clients again and again. Meanwhile, the GEFMA certificate is often required as a basic requirement for participation in CAFM tenders.

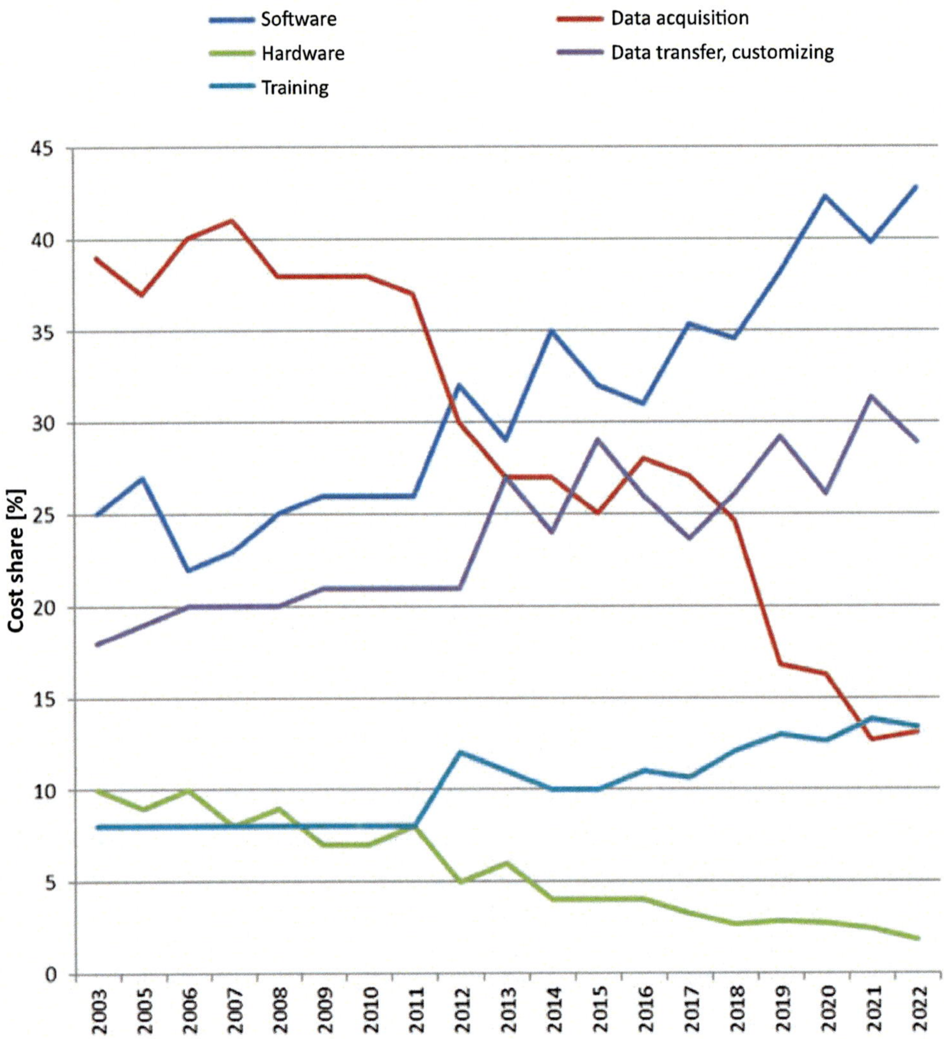

Fig. 3.3 Development of costs of a CAFM implementation

3.3 Benefits of BIM for Facility Managers

3.3.1 Why BIM for FM?

Should facility managers be interested in BIM? In (Teicholz 2013) this is justified as follows:

> "An inappropriately large amount of time is spent locating and checking specific object and project information from previous activities."

This hampers efficient FM processes and decisions. For some time now, it has been possible to observe how the construction industry uses digital methods in the design and construction of buildings and facilities. Siemens Real Estate (SRE) defines BIM in its own standard (NN 2017b) as follows:

> "Building Information Management (BIM) is a model-based, digital, interdisciplinary and verifiable working methodology. Virtual, three-dimensional building models are linked with non-geometric, alphanumeric data, providing a consistent image of all planning, execution and operator-relevant information. These multi-dimensional building models are the digital and visual representation of the building, the physical and functional properties and their relationships to each other. The different data and information of the project participants enable an accurate and verified transfer of 'as built' data into operation."

In design and construction, BIM plays an important role in establishing digital, collaborative processes. This enables all parties involved to design and construct buildings in a shorter time, with higher quality and at lower costs. It has long been known that the majority of later construction and management costs are already determined in the planning phase. Therefore, facility managers should actively participate in the design and construction phases, whereby their competence e.g. in maintenance and operation is already taken into account in the design process and can thus lead to sustainable added value in the life cycle of buildings. This also applies to refurbishment or renovation projects. It is assumed that facility managers can benefit from BIM to a great extent. By using BIM, it is possible to reduce costs in the lifecycle and increase the value of buildings (Ashworth et al. 2016). This applies in the same way to buildings, infrastructure and assets.

BIM models support facility managers in improving processes and reducing associated costs. Simulations support this development, e.g. by crowd simulation of building users anticipating evacuation, or by simulation of space occupancy, safety, cleaning or energy consumption, whereby the BIM model is always the data basis of the simulation (cf. Sect. 2.11).

3.3.2 Benefits of BIM in Commissioning

When lifecycle phases of buildings are considered, it becomes apparent that the highest loss of information occurs during the transition from construction to the start of use (cf. Fig. 3.4, based on Sacks et al. 2018). Board out means the traditional design based on a variety of 2D CAD drawings.

In particular, when transitioning from construction to operation, facility managers want to take over relevant building information in order to ensure efficient commissioning and appropriate building operation. Since BIM models are increasingly available, it is of major interest to take over, use and maintain data from the construction phase seamlessly. This leads to higher quality of commissioning with less effort, saving a considerable amount of time and money.

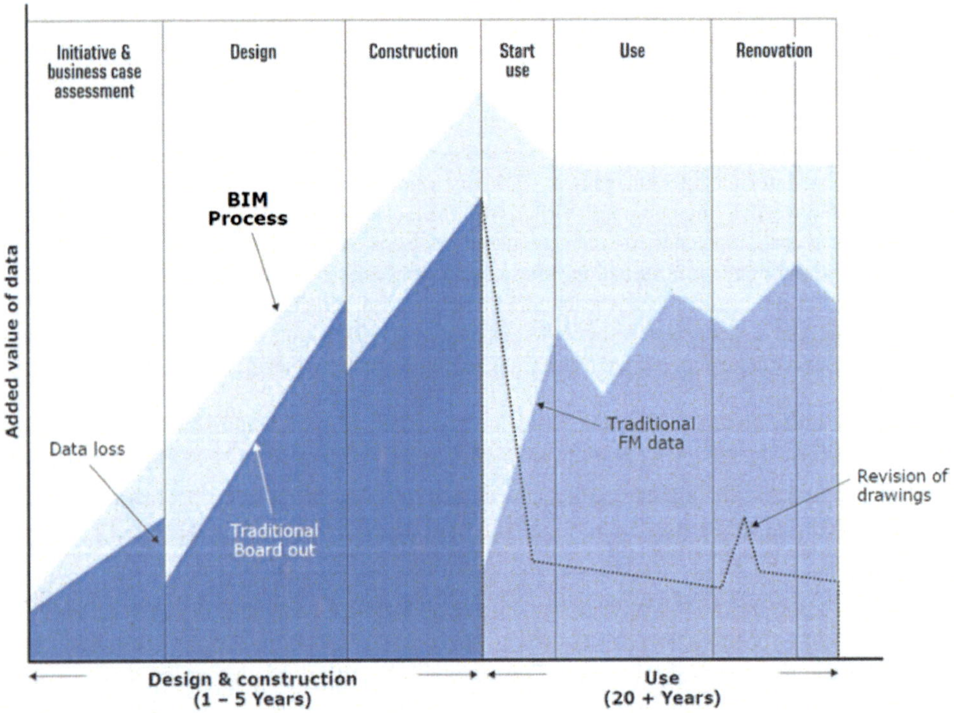

Fig. 3.4 Data loss during transition of project phases with and without BIM

The preparation of multi-year forecasts for FM processes (e.g. in maintenance) is based on the availability of suitable building and asset data. If the data required is provided in BIM models, this allows facility managers, for example, to plan and carry out maintenance work faster and more reliably, using CAFM or computerized maintenance management systems (CMMS). Similar benefits can be seen in warranty management.

The ability to use data relevant to operation from BIM models during the transition from construction to operation phase represents a high potential for benefits in FM and is therefore an essential aspect of the economic viability of BIM in FM. This eliminates duplication of data.

3.3.3 Benefits of BIM in Operation

In the case of commissioning (see Sect. 3.3.2), only the transfer of data from the BIM model to IT systems that are used for operation and maintenance of buildings is required. One could argue that digital building models become obsolete thereafter.

The most important question related to BIM for facility managers is: Is there a Business Case that justifies updating and maintaining of BIM models for the remaining lifecycle of buildings, i.e. for the operational and decommissioning phase?

This is not a trivial question, because the decision to maintain and update BIM models during the lifecycle has consequences for the information management of real estate and facility management. This causes additional costs.

If it is decided to use BIM models during operation, this requires appropriate IT environments that enable software users or systems to access the models in real time and to visualize and change them if necessary. This affects both graphical-geometric and alphanumeric data. Such IT environments are referred to as BIM server platforms (see Chap. 4). The use of such platforms requires that the data contained in the BIM models be kept up to date.

During operation, for example, conversions, renovations or maintenance work is carried out. The exchange of technical equipment and components (e.g. pumps, fans) must be reflected in the master data, regardless of whether the BIM model was used during the maintenance process or not. For this purpose, reliable interoperability between the BIM systems on the one hand and the CAFM systems on the other hand is indispensable. This also causes effort and costs. Likewise, BIM authoring tools must also be usable during the operational phase. In addition, a robust management of multiple model versions must be ensured.

For the acceptance of BIM models the usage must ensure cost savings and/or quality improvements in operations. The amount of the cost savings must exceed the costs of creating and maintaining the models. So far, no reliable economic data (cf. Chap. 6) seem to exist for such business cases, primarily because using BIM in operation is still in its infancy.

As practical experience is gained, figures on profitability of BIM will also be available, similar to what is known from CAFM projects. In addition to quantifiable aspects, there are also qualitative benefits which drive the implementation of BIM during operations. For such economic considerations, GEFMA guideline 460 (NN 2016a) is recommended, even though it is about the economic efficiency of CAFM projects.

3.3.4 Benefits of BIM in Renovation and Conversion

BIM has already proven itself in the conception, planning and construction of buildings. There is also a great deal of specialist literature on this subject (Borrmann et al. 2021). Reference is also made to BIM information and publications from organizations such as the international buildingSMART alliance (NN 2021d).

During the lifecycle, buildings are subject to a number of renovations (retrofit) and conversions. Here, BIM offers exactly the same benefits as in conception, planning and construction: higher quality at lower costs and shorter implementation time. In addition, BIM can be helpful in scenario planning.

In order to successfully support renovation and conversion processes using BIM, BIM models are needed that provide architects and construction companies with appropriate and up-to-date information for planning and implementing including changes to a building.

An example will be explained hereafter. An existing building was designed based on a traditional cell office design. It is to be redesigned to enable new ways of working, possibly also under pandemic conditions. This is a modernization/conversion project. Networking of employees, extended collaboration options, team work and informal work styles as well as additional hygiene regulations are more important in the new concept. The interior design must be attractive and comfortable for this purpose, so that employees can carry out their activities effectively. A great user experience is the major focus. The project includes a fundamental redesign of floor plans, but also the design of the ventilation and air conditioning systems, in order to guarantee an air quality adapted to the new room design. With the help of BIM tools, an (IFC) model of the existing building is created and forwarded to the potential contractors, so that they can plan the conversion and estimate the associated costs based on the models. The processed BIM models which are completed and returned to the client provide a detailed insight of ideas and approaches proposed by the different contractors. It the end, this enables the client to make well-founded decisions regarding concepts and contractor selection.

3.3.5 Benefits of BIM in Everyday Work

In what ways can BIM provide benefits for everyday work during the operational phase?

A short answer to this question would be: BIM models enable fundamentally new ways to analyze buildings and to better understand their operation. In consequence, this improved understanding will enable facility managers to provide their services with higher quality and increased efficiency. To explain this, the advantageous characteristics of BIM models for tasks in building operations must be explained.

A fundamental aspect of BIM models is the ability to describe the geometry of buildings and technical assets in a comprehensible way. The ability of humans to easily understand visual information is decisive here. For example, if you wanted to describe the design of a building in detail, such as shown in Fig. 3.5, you would probably need hundreds of pages of text and tables, and even then the description would probably still be incomplete or not detailed enough.

BIM models enable a quick and accurate analysis of condition and properties of their objects. Maintenance activities, for example, benefit from this improved understanding, which can accelerate the preparation of work processes.

In the same way, BIM models can support space management by providing a deeper understanding of the spatial features of buildings. This, for example, enables better and faster decisions to be made in space planning. Basically, a BIM model with sufficient data can be used to estimate the expected energy, maintenance, cleaning and space costs.

Fig. 3.5 Example of a BIM model

3.3.6 The Digital Twin Principle

In the field of business analytics, BIM models also provide improved insights and support of decision-making for real estate and facility managers.

A Digital Twin is a virtual representation of a real, physical object that also provides information about its current state using sensors and IoT technology. Often, sensors are attached to real objects within the building to transmit corresponding state data. The simulation capability of the physical building is often also considered an important characteristic of a digital twin.

This digital twin principle was developed at the beginning of this millennium in the context of Product Lifecycle Management (PLM) in the automotive industry. It quickly found its way into the manufacturing sector (especially mechanical engineering and electrical engineering). A well-known example is from General Electric (GE), where the model of an engine or a wind turbine is linked with current data from its operation. The presentation of operating parameters in geometric models offers new possibilities to better understand the state of objects and, if necessary, to intervene for correction. The analysis of potential risks is also made easier and effects of incidents such as accidents can be predicted more quickly and reliably. This approach is called analytics-based maintenance. In the example of GE, the condition of wearing parts of engines are presented together with their operating parameters in the model. Engineers are thus able to analyze the failure risks of the parts and take appropriate measures in time.

The same principle applies to the operation of buildings. BIM models can be linked with relevant operational and other data and allow facility managers or other stakeholders to assess the conditions and potential failure risks of building or technical components in a very intuitive way.

Digital twins are also a good application area for successful integration of different technologies such as BIM, CAFM and IoT. There are numerous interesting use cases, e.g. in occupancy management, energy and environmental management, security and fire protection, up to various facility services. Fig. 3.6 shows such a constellation, in which different providers cooperate within a digital twin.

3.3.7 Requirements for the Use of BIM Models in Real Estate Operations

It is very important to realize that the use of BIM models over a period of many years has important consequences for information management. Therefore, it is recommended to involve the IT department in early stages of a BIM project.

Some practical aspects when using BIM models are:

- *BIM-related document management*
  BIM models are software objects. This means that you have to choose certain notation forms (e.g. IFC or vendor-specific formats). It also has to be clarified how the different model versions are to be managed.
- *Interoperability with other systems*
  CAFM and CMMS systems have to be linked with BIM models so that they are accessible for facility managers. In addition, these systems contain data that must also be available in the BIM models. For example, if a defective component such as a pump or fan is replaced, the component data generated during procurement and

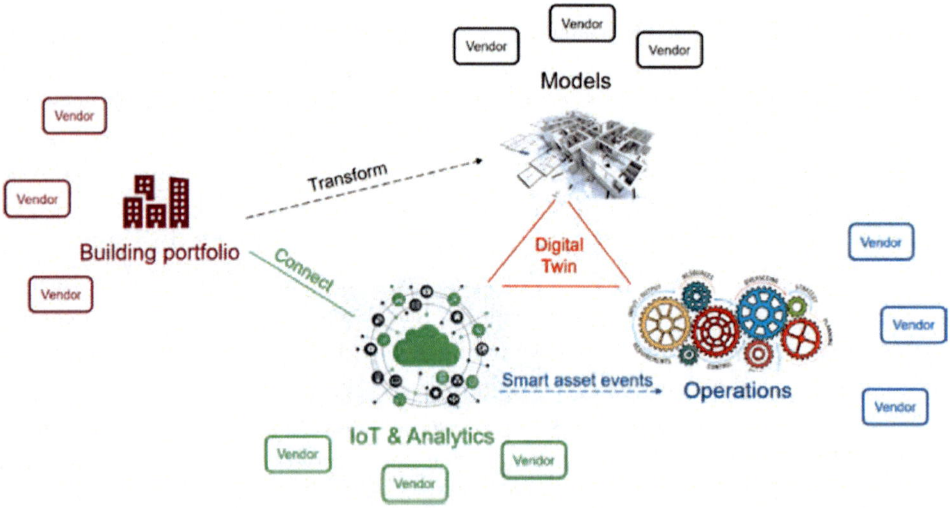

Fig. 3.6 Interaction of BIM, CAFM and IoT in a digital twin

exchange are typically processed in these systems and must be managed in the asset inventory system of the building.

- *Security and authorization*
  If BIM models are an integral part of the information architecture, data access must be planned and managed. In most cases, the historization of data must also be tracked in an audit trail so that it can be checked who changed which data and when. In addition, the data must be protected against unauthorized use and manipulation. These are typical IT and information management topics that must also be taken into account in BIM projects.
- *BIM competence development and assurance*
  This important task can be solved by suitable training both in-house and externally.

3.4 BIM Basics for Facility Managers

Facility managers must deal with the BIM methodology to varying degrees depending on their role in the company. They can be involved in a BIM project team of responsible planners and construction companies as a representative of the operational phase or as the responsible person for corporate real estate management in order to implement BIM in the company and or in specific projects.

Regardless of the role of the facility manager in a BIM project, FM has to describe in detail the information requirements and deliverables of the project participants. This has to be implemented as a fixed component in the BIM projects.

On the way to defining the requirement definition a specific BIM strategy and the goals pursued with it must be formulated (see Fig. 3.7). The recommendation in line with ISO 19650 is that the starting point in a BIM project is the organizational information requirements (OIR). These are the information requirements that organizations need to achieve their strategic objectives and manage their day to day activities. Typically these are represented by management reporting requirements across the organization, e.g., health and safety, asset management, warranties etc. From the OIR specific

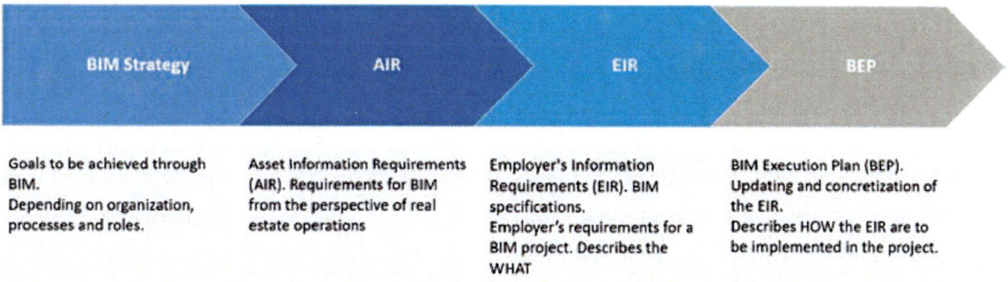

Fig. 3.7 BIM requirement documents

asset information requirements (AIR) can be identified for BIM. They deal with information specific to managing assets for the BIM project and can be derived from the perspective of real estate operations. These requirements must be anchored in the respective BIM projects. For this purpose, the contents of the AIR must be transferred to the BIM employer information requirements (EIR) and BIM execution plan (BEP) (see also Sect. 7.2).

In order to formulate the requirements for BIM or the BIM project in a targeted and binding manner and to implement them in a project, the competence of external BIM experts (e.g. BIM consultants or BIM information manager) is often used. However, a basic understanding of the BIM method and its terminology is essential for facility managers. First, essential aspects of the BIM method and its terminology will be briefly explained before the contents of the AIR is presented in detail.

3.4.1 BIM Definition

As defined in Sect. 2.2, BIM is a method or process in which a 3D building information model (geometry and semantics) is gradually created during a BIM project, which is handed over to FM at the end of the project (FM handover).

The building information model then provides a valid data basis for digital building operation, for example with a CAFM system. In addition to the geometric 3D model, the alphanumeric object information is important for FM, as this is used to map digital processes (e.g. maintenance) in a CAFM system.

3.4.2 BIM Maturity Model

The BIM Maturity Model, see Fig. 3.8, based on Borrmann et al. 2021 is used for developing implementation plans for BIM and comprises four maturity levels (level 0–3), which show how comprehensively the BIM method is used in a project or organization. The stages also mark milestones in the implementation of a national BIM strategy.

3.4.3 BIM Dimensions

BIM models are constructed as multi-dimensional (multi-perspective) information models that can contain additional layers of information beyond the geometric dimensions, which allow architects and engineers, for example, to create a model-based schedule and cost plan. In the context of BIM, five additional dimensions (3D to 7D)

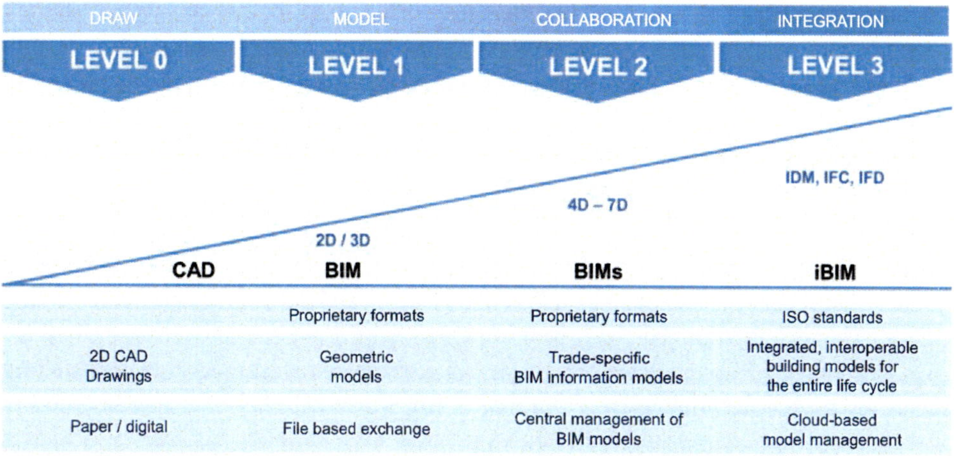

Fig. 3.8 BIM maturity model

beyond the geometric 2D representation are usually distinguished (see Fig. 3.9, based on NN 2021ax).

3.4.4 Open and Closed BIM

Various project participants (e.g. architect, structural engineer, MEP, FM) are involved in the creation of the digital building information model, who rely on so-called BIM

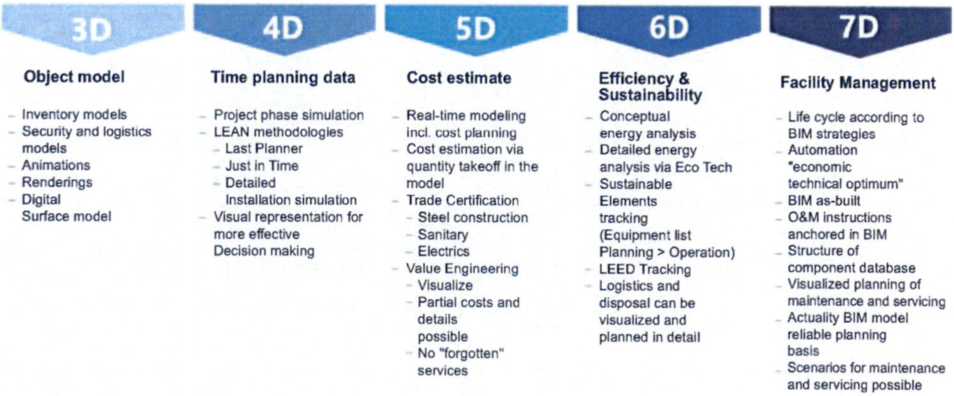

Fig. 3.9 BIM dimensions

authoring tools (cf. Sect. 4.2.1) for modeling. If all project participants use the same BIM authoring tool, the model-based exchange can take place based on the vendor-specific, proprietary data format of the authoring tool used. This is called Closed BIM. This also applies if several tools of a software suite of one vendor are used with their proprietary data formats.

However, architects, structural engineers and MEP engineers usually use BIM authoring tools from different manufacturers, which are optimized for the respective planning task. In order to still ensure a smooth and loss-free data exchange in the project, open, standardized data formats such as IFC (cf. Sect. 3.4.5 and 5.3.2) are used. When working in this way in a BIM project, one speaks of Open BIM.

Whether only one company or all project participants work open or closed is indicated by the prefix "little" or "big" (cf. Fig. 3.10).

The following combinations are possible:

- *Little Closed BIM*
  Use of the BIM method in a company based on proprietary data formats for internal data exchange.

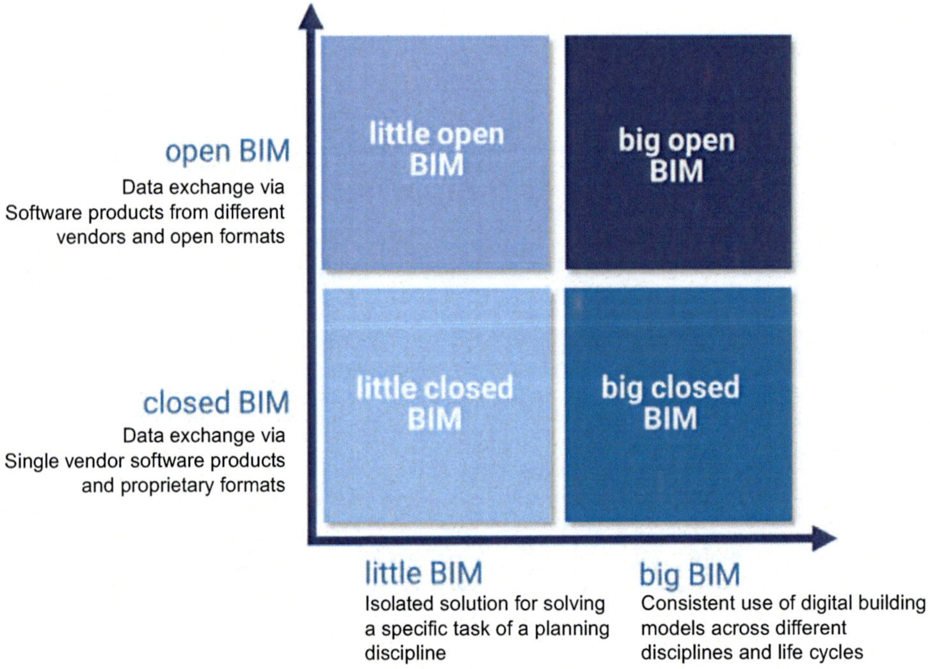

Fig. 3.10 Overview open/closed, little/big BIM

- *Big Closed BIM*
  Use of the BIM method over several lifecycle phases of a building with model-based data exchange between the participating companies based on proprietary data formats, mostly of a vendor of a BIM suite.
- *Little Open BIM*
  Use of the BIM method in a company, whereby software products from different manufacturers realize data exchange via open data formats (e.g. IFC).
- *Big Open BIM*
  Use of the BIM method over several lifecycle phases of a building with model-based data exchange between the participating companies based on open data formats (e.g. IFC).

Further information on term definitions can be found in Helmus et al. (2019).

3.4.5 BIM Discipline Models and the CDE

A key tool in BIM projects are BIM authoring tools, which are used for geometric component modeling and for the definition of semantics. The use of the BIM method with simultaneous access of all disciplines to a common BIM model has not been established in practice, even in BIM projects of the BIM maturity level 3 (see Sect. 3.4.2) due to organizational and technical restrictions (Borrmann et al. 2019b).

Rather, BIM projects are based on federated discipline-specific models (object planning, structural planning, MEP, FM) that are regularly combined into a coordination model. Coordination models enable discussions between selected disciplines and their discipline models.

The creation of the coordination models and thus merging selected discipline models is done by the BIM project coordinator, whereas checking and release is done by the BIM manager. As a platform for managing and versioning models, a common data environment (CDE) must be established in the project (cf. Sect. 4.3). Coordination models are created in the project at defined milestones (e.g. at the end of a planning/performance phase). Based on a coordination model, collision tests or cross-disciplinary evaluations can take place and the degree of completion can be checked according to BEP. Changes are not made within the coordination model itself, but are continued in the original discipline models. The coordination models are versioned and stored in the CDE.

The as-built model is created as the final model at project completion. This represents a checked, digital version of the actual building. For operations, a reduced operating model is derived from the as-built model. It only contains the operationally relevant content. The operating model is transferred to the IT systems used during operations. In particular, the alphanumeric asset information is transferred to a CAFM system to be used and updated during the operational phase (cf. Sect. 4.3.2). In case of major construction or renovation measures during operations, current operational data can be used to provide as-built models for the renovation process.

3.4.6 Open Data Exchange of BIM Models

The correct and automated exchange of digital building information models is an essential goal of BIM projects. As a vendor-independent, open format, the IFC schema (Industry Foundation Classes) developed by buildingSMART has become the leading open BIM standard internationally (see Sect. 5.3.2). In version 4, the IFC format is recognized as ISO Standard 16739 (NN 2018a).

Many of the BIM authoring tools and CAFM systems established on the market now support the IFC format (import, export). Basic requirements for data exchange or for the integration of BIM models into a CAFM system are described in catalog A15 of the GEFMA guideline 444 (NN 2020a) and are checked for compliance during CAFM certification.

IFC can be seen as a data schema according to which data exchange between two systems takes place. An IFC file can be regarded as a copy of the model, which contains both geometric and semantic object information. It can be used for data exchange of text/graphic data in a similar way to a PDF document. Objects are not edited within an IFC file in general, but changes are made in the original source model instead and then exported to an IFC model again. Based on the IFC schema, a building is mapped according to a uniform structure (see Fig. 3.11).

In addition to the uniform structure, predefined property sets (IFC Common Property Sets) are available for object description. However, these predefined property sets often do not suffice to describe objects according to the requirements from an FM perspective. In order to meet this requirement, so-called User-defined Property Sets can be defined additionally within the IFC schema. This offers enough flexibility to meet special requirements at the client or project level.

A legally binding definition of property sets required for real estate operations is carried out in the asset information requirements (AIR) and employer information requirements (EIR).

Fig. 3.11 IFC building structure

3.4.7 Definition of Model Contents

The 3D building information model is an important deliverable within BIM projects in terms of a data drop. How objects are modeled in the BIM authoring tools and which alphanumeric or geometric information is captured depends on the respective BIM application case (see Sect. 3.3 and Chap. 6) and is described in detail in the BIM documents EIR and BEPvia the level of development. The level of development (LOD) essentially describes which information is to be provided by the project participants at a certain point in time (e.g. at the end of a service phase in the project). It includes the level of geometry of BIM objects (Level of Geometry—LOG) mapped in the model and their descriptive alphanumeric information (Level of Information—LOI). With the LOI, the majority of semantic information of the digital building model is specified by attributes. Further information can be found in Borrmann et al. (2019a, b).

The definition of the elaboration degree is specified in five levels of development. LOD 100 describes the lowest level and LOD 500 the highest (cf. Fig. 3.12). LPH's indicate the service phases within a construction project according to HOAI. The levels of development show how the building information model is gradually becoming more detailed or enriched with further information over the course of the project in the respective service phases.

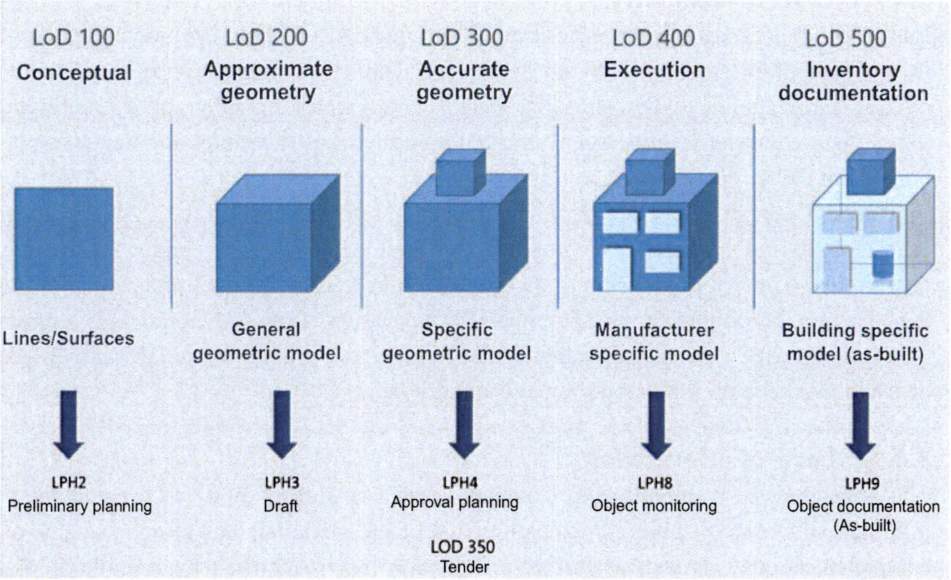

Fig. 3.12 Overview of the levels of development (LODs) in a BIM project

- *Level 100*
  Conceptual representations and studies
- *Level 200*
  Information on dimensions and size of relevant building elements as well as their relationships with each other
- *Level 300*
  Basis for implementation: tendering-ready information with specifications
- *Level 400*
  Manufacturing-ready execution planning
- *Level 500*
  Documentation of the executed object (as-built)

From an FM perspective, in particular the highest level of detail (LOD 500) is of interest, as this describes the final state of the building (as-built), on the basis of which the FM handover takes place.

In order to achieve this goal, it is essential to specify the requirements from the FM perspective (AIR) in BIM project documents EIR and BEP in a binding manner. Furthermore, the model quality and the model progress must be checked in each service phase or at each milestone. Only if the digital building information model is continuously quality-assured and updated over the entire project course, a valid as-built documentation at the end of the project can be expected.

3.4.7.1 Level of Geometry

Basically, the increasing geometric level of detail (LOG) of the BIM model elements—from a symbolic to a simplified and finally to a detailed representation in the discipline models—corresponds to the increasing scale accuracy in traditional graphical representations. The geometric level of detail is usually aligned to the progressing depth in planning along HOAI phases.

The demand for LOG from the FM perspective also depends on the CAFM software used in facility management. Requirements of the CAFM software vendor for the interface to the geometry model should therefore be reflected in the requirements definition. If necessary, in addition to the 3D model, 2D derivations from the model (e.g. sections) are required, which are linked to the CAFM system as a graphic basis. This should also be taken into account in the requirements definition.

3.4.7.2 Level of Information

The alphanumeric (semantic) level of detail (level of information—LOI) includes the information for the unambiguous identification of the model elements such as name and component number as well as further information that is required for specific applications or in certain service phases.

The alphanumeric information is of particular importance for the FM, as this represents the data basis for the digital processes in a CAFM system (e.g. in space, maintenance or fault management).

The following points must be considered in particular when defining the LOI:

- Use of a consistent and standardized classification of the objects to be described (e.g. according to CAFM-Connect or similar classification systems),
- Definition of the contents required for processes and objects in the form of property sets.

Standards such as CAFM-Connect (see Sect. 5.3.5) provide both the framework for a uniform object classification and so-called BIM profiles, which provide the required data content and the transfer format of the alphanumeric content to be delivered on the basis of IFC (IFCXML and IFCZIP) (Fig. 3.13).

The documentation of the LOI is often still carried out analogously in the form of a matrix in the BIM documents. The list usually describes the following content:

- For which application case,
- To which objects,
- Which properties,
- In which format,
- At what time,
- To be captured by whom.

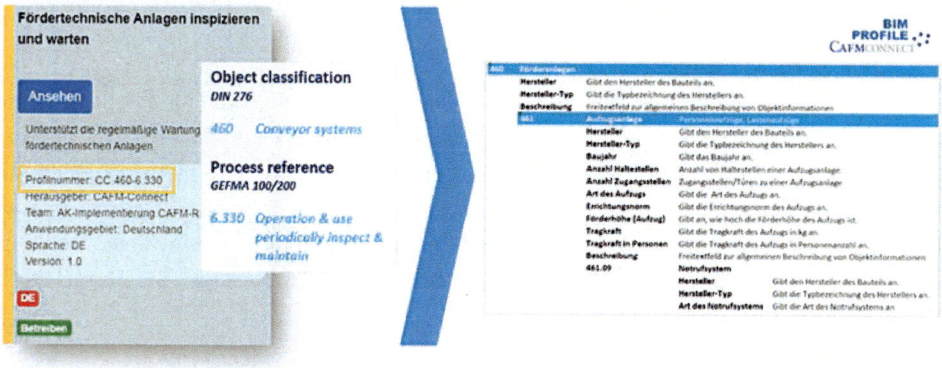

Fig. 3.13 Description of property sets based on CAFM-Connect

Meanwhile, there are also IT systems on the market that support the creation and updating of digital EIR's. With such systems, the property sets to be delivered in the project can also be described and distributed digitally.

3.4.8 Asset Information Requirements

As already mentioned, the requirements for a BIM project must be specified clearly and fixed in the project right from the start. From the perspective of real estate management, this requirement definition is carried out in the asset information requirements (AIR).

The AIR are usually created project-wide within the organization and are to be transferred to the BIM documents EIR and BEP or reference is to be made to them. The contents of the AIR is analogous to the contents of the EIR (cf. Sect. 7.2). However, they are described exclusively from the perspective of FM and have no concrete project reference. A possible structure of the AIR is shown in the following table of contents:

1. BIM applications
2. Digital basis provided
3. Digital deliverables
4. Organization and roles
5. Collaboration strategy
6. Delivery times
7. Quality assurance
8. Model structure and model content
9. Technology

The most important items are described in more detail below.

3.4.8.1 BIM Use Cases

First, the applications and business processes relevant from an FM perspective need to be described for a targeted requirement definition. These are usually the creation of the operational documents. However, the contents of the operational documents also depends on the relevant operational processes, i.e. different content is required for maintenance and for space management. This differentiated view needs to be taken into account in the requirement definition in order to be able to describe the correct digital deliverables.

3.4.8.2 Digital Deliverables

As part of its services, the contractor is to create, check and transfer digital deliverables to the client with a focus on FM. The deliverables are described in terms of service phases and the specification of the LOD. Deliverables are files that are to be handed over to the client at the end of a service phase. Deliverables can, for example, be models, 2D plans derived from the model, semantic information or file documents. The level of detail (LOG, LOI) of the digital models to be delivered needs to be specified.

3.4.8.3 Collaboration Strategy

This is where the process of collaboration needs to be described in general. What is the role of FM represented by the BIM information manager in the BIM project, and what tasks are associated with it (see Sect. 7.2)?

3.4.8.4 Quality Assurance

As already stated in Sect. 3.4.7, continuous quality assurance in a BIM project is a key component for valid as-built documentation. Quality assurance of the requested digital deliverables is generally the responsibility of the contractor. From an FM perspective, at least the following content must be checked:

- Compliance with the given model structure and model content,
- Compliance with the given data formats,
- Compliance with the appropriateness of data size,
- Consistency of derived plans and digital models,
- Consistency of models with reality.

FM is to be involved in the testing by an information manager.

3.4.8.5 Model Structure and Model Content

Naming, classification, structure and organization of digital models are essential for the use of models by the employer. The focus here is always on FM. The contractor has to ensure compliance with the specified modeling conventions of digital deliverables:

- Coordinate systems,
- Units,
- Structuring (e.g. asset identification key),
- Classification systems (e.g. CAFM-Connect),
- Elaboration levels (LOD, LOG, LOI),
- Modeling regulations.

3.4.8.6 Technology

In this section, information is provided about the common data environment (CDE) as well as about data exchange formats. The definition of exchange formats (e.g. IFC, CAFM-Connect, COBie) depends on the IT systems of the CDE and the once used in the operational phase. Furthermore, the process of data integration after the construction phase should be described.

3.5 Integrated Digital Delivery—An International Approach in BIM Projects

As BIM systems mature, there is growing interest in their use beyond the planning and construction phases. However, effective implementation on a large (international and partly also national) scale has not yet been achieved. However, there are developments that lead to the structural implementation of the BIM method up to the use for existing buildings during the entire lifecycle. It is noteworthy that some governments initiate and promote the use of BIM models and modeling tools in the lifecycle at the national level. Examples of proactive governments are Great Britain and Singapore. With the "Integrated Digital Delivery" (IDD) of the Building and Construction Authority Singapore (NN 2021n), an interesting and much-noticed approach was published. This approach includes some good procedures adopted in numerous BIM guidelines and is therefore described in this section.

3.5.1 Integrated Digital Delivery

The core idea of IDD is to minimize disruptions during the construction of buildings, accelerate the actual construction phases and facilitate the transfer to the FM teams so that projects can be completed faster (cf. Fig. 3.14). IDD therefore includes the use of digital technologies to integrate work processes with all those involved in the lifecycle of a building.

In addition to the CAFM/IWMS systems used in the operation of buildings throughout the lifecycle, BIM tool sets are in the focus of implementing this digital approach.

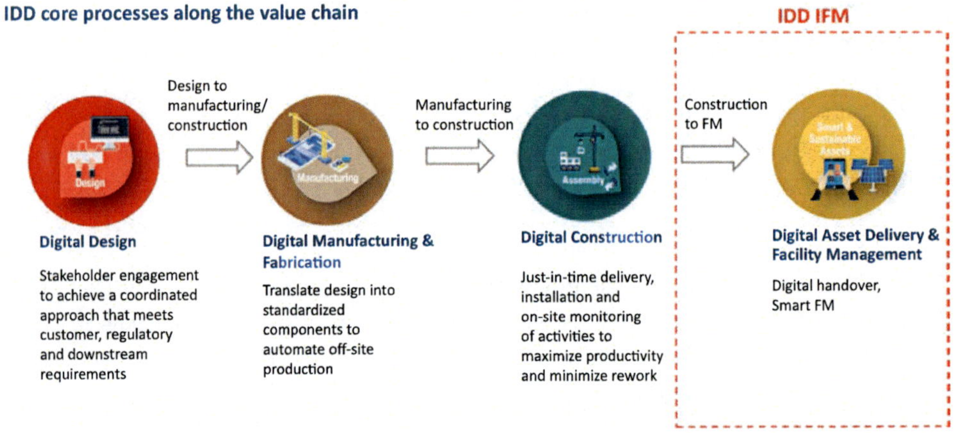

Fig. 3.14 The four phases of integrated digital delivery (IDD) of buildings

Experience shows that the IDD approach only works if some changes are made to the BIM models during digital data transmission (model transformation) so that they are suitable for operations in the next stage (see Fig. 3.15).

This is done in the commissioning phase of buildings, in which the BIM models are adapted so that data that is only of interest for planning and construction are removed. Furthermore, data is added that is of interest for space and facility management, as it is an essential part of the operational phase.

The BIM models handed over must be constantly updated during the use phase of buildings always representing the actual state (as-is). If assets are replaced one-to-one, this can usually be done easily with the CAFM systems in conjunction with existing models. Geometric changes, such as changes to floor plans, usually introduce new elements into the building. Here, the transformation process usually has to be carried out again to update the changes made.

3.5.2 The Potential Role of FM in the IDD Process

Facility managers must assume a new role in such processes. This key role must be proactive (cf. Fig. 3.16), with early involvement being an important aspect.

The engagement of the FM ranges from an advisory role in the early phase to a leading role in the commissioning and operational phase.

In terms of information systems, this type of digital approach requires the easiest possible data exchange between all systems used during building operations. These include BIM models and tool sets, CAFM/IWMS systems, but also BMS and IoT systems.

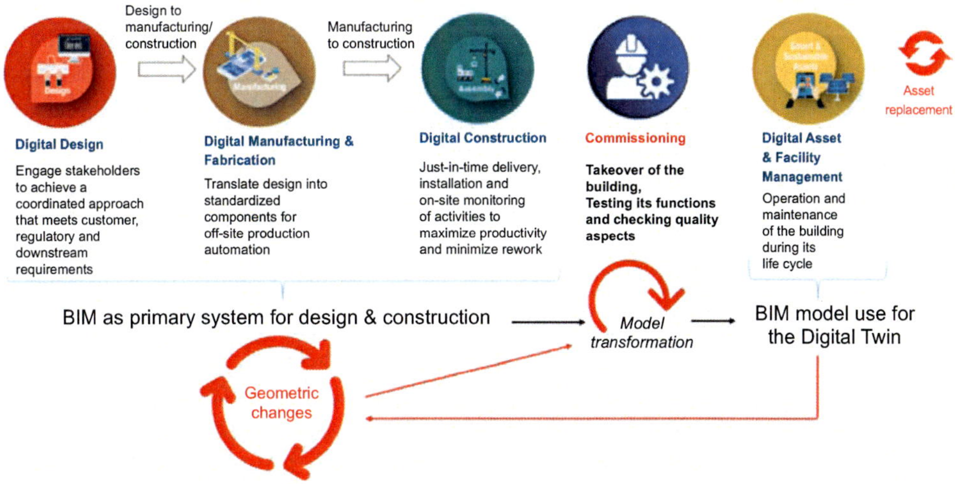

Fig. 3.15 Management of BIM models during the lifecycle

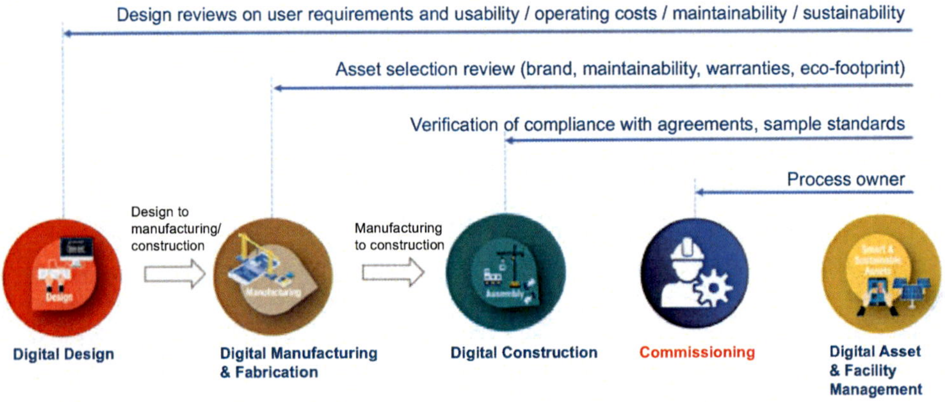

Fig. 3.16 The potential role of FM in the IDD process

To achieve this, the concept of a common data environment mentioned in Sect. 3.4.5 and elaborated in Sect. 4.3 is of importance. A CDE is used as a single data repository that is used to collect, distribute, and manage the required documentation.

Conceptually, this approach fits perfectly with the integrated digital delivery (IDD) model. However, the problem is that software vendors today produce different and only partially interoperable CDE solutions. This current CDE diversity and the lack of standardization of CDE initiatives often still represent a barrier to users in successfully utilizing this approach.

3.6 Summary

Chapter 3 presents the basic principles required of the facility manager for a successful use of the BIM method from an FM perspective.

The starting point for this task is first the explanation of the evolution of CAD-based approaches to today's understanding of BIM. It becomes clear that reservations with regard to the BIM method already existed in a similar way in the past when CAD was introduced and could be overcome.

CAFM systems today play a key role in building operations and are an integral part of FM organizations both on the client and contractor side. In order to understand the importance of CAFM systems with regards to the BIM method in the future, Sect. 3.2 first deals with important CAFM basics. The CAFM understanding, the market of CAFM software, their scope of functions and typical challenges of CAFM projects in relation to the BIM method are explained.

Section 3.3 is dedicated to the central question of the benefits of BIM for facility managers and building operations. For this purpose, important BIM use cases are considered along the different processes of the operational phase, from commissioning, operation, to renovation and conversion. This includes the consideration of economic, technical and above all organizational aspects as well as the benefits of BIM in everyday business. With the explanation of the principle of digital twins, the concepts of BIM, CAFM and IoT are linked together in order to use BIM models not only for the static description of buildings, but also in the context of dynamic behavior of buildings in operation. In this way, with digital twins, a new quality of decision making in operations based on reliable predictions of building behavior with the aim of improving energy efficiency and sustainability becomes achievable.

The implementation of the BIM method in FM, but also the understanding of BIM on the part of planning and construction for FM and vice versa is dealt with in the following section. For this purpose, fundamental definitions of the BIM method are presented as well as different stages for the implementation of BIM using the BIM maturity model and typical additional dimensions of BIM for the consideration of time, cost and the operational view are presented.

In order to implement this practically, however, important basics of model-based data exchange must be considereded. Different forms of using open or vendor-specific (closed) data formats for collaboration with BIM are explained and the handling of discipline and collaboration models from the FM perspective is shown. Finally, the representation of BIM basics leads to practical advice on the definition of asset information requirements (AIR) as a specification of requirements for BIM from an FM perspective.

The chapter closes with the presentation of the internationally recognized approach to "Integrated Digital Delivery" of the Building and Construction Agency from Singapore and thus presents international experiences in the introduction of BIM also for the operating phase of buildings.

References

Ashworth S, Tucker M, Druhmann C, Kassem M (2016) Integration of FM expertise and end user needs in the BIM process using the Employer`s Information Requirements (EIR), May 2016

Borrmann A, Elixmann R Eschenbruch K, Forster C, Hausknecht K, Hecker D, Hochmuth M, Klempin C, Kluge M, König M, Liebich T, Schöferhoff G, Schmidt I, Trzechiak M, Tulke J, Vilgertshofer S, Wagner B (2019a) Leitfaden und Muster für den BIM-Abwicklungsplan. Publikationen BIM4INFRA 2020, Teil 3

Borrmann A, Elixmann R Eschenbruch K, Forster C, Hausknecht K, Hecker D, Hochmuth M, Klempin C, Kluge M, König M, Liebich T, Schöferhoff G, Schmidt I, Trzechiak M, Tulke J, Vilgertshofer S, Wagner B (2019b) Handreichung BIM-Fachmodelle und Ausarbeitungsgrad. Publikationen BIM4INFRA 2020, Teil 7

Borrmann A, König M, Koch C, Beetz J (Eds.) (2021) Building Information Modeling – Technologische Grundlagen und industrielle Praxis. 2. aktualisierte Auflage. Springer Vieweg, 2021, 871 S

Helmus M, Meins-Becker A, Agnes K, et al. (2019) TEIL 1: Grundlagenbericht Building Information Modeling und Prozesse. Forschungsbericht Bergische Universität Wuppertal. Fakultät für Architektur und Bauingenieurwesen. Lehr- und Forschungsgebiet Baubetrieb und Bauwirtschaft

Kalweit T, May M (2017) Cloud-Technologie im Facility Management. In: Bernhold T, May M, Mehlis J (Eds.): Handbuch Facility Management. ecomed-Storck GmbH, Landsberg am Lech, 55. Ergänzungslieferung, Dezember 2017, 24 S

May M (Ed.) (2018a) CAFM-Handbuch – Digitalisierung im Facility Management erfolgreich einsetzen. 4. ed., Springer Vieweg, Wiesbaden, 2018, 713 p

May M (2021) 20 Jahre GEFMA-Arbeitskreis Digitalisierung – Mehr als nur CAFM und Richtlinienarbeit für das FM. Facility Management 27(2021)1, 44–47

NN (2016a) GEFMA Richtlinie 460: Wirtschaftlichkeit von CAFM-Systemen, Mai 2016, 27 S

NN (2017a) GEFMA Richtlinie 420: Einführung von CAFM-Systemen, Juli 2017, 7 S

NN (2017b) BIM@Siemens Real Estate, Standard Version 2.0 vom 25.10.2017. https://assets.new.siemens.com/siemens/assets/api/uuid:caceb1c2b181de452d5f9ec00b1cb0d1242d5498/version:1520000392/bim-standard-siemens-real-estate-version-2-0-en.pdf (retrieved: 25.04.2021)

NN (2018a) ISO 16739-1: Industry Foundation Classes (IFC) for data sharing in the construction and facility management industries Part 1: Data schema. International Organization for Standardization, 2018-11

NN (2020a) GEFMA Richtlinie 444: Zertifizierung von CAFM-Softwareprodukten. Februar 2020, 21 S

NN (2021b) Marktübersicht CAFM-Software. GEFMA 940, Sonderausgabe von „Der Facility Manager", FORUM Zeitschriften und Spezialmedien GmbH, Merching, 2021, 198 S

NN (2021c) GEFMA Richtlinie 610: Facility Management-Studiengänge. 2021-11, 3 S

NN (2021d) https://www.buildingsmart.de/ (retrieved: 26.04.2021)

NN (2021n) https://www1.bca.gov.sg/buildsg/digitalisation/integrated-digital-delivery-idd (retrieved: 16.08.2021)

NN (2021ax) https://h-m-consult.com/ (retrieved: 06.12.2021)

NN (2022b) Marktübersicht CAFM-Software. GEFMA 940, Sonderausgabe von "Der Facility Manager", FORUM Zeitschriften und Spezialmedien GmbH, Merching, 2022, 202 S

Sacks R, Eastman C, Lee G, Teicholz P (2018) BIM Handbook. 3rd ed., John Wiley & Sons, Hoboken, New Jersey, 2018, 659 S

Teicholz P (Ed.) (2013) BIM for Facility Managers. John Wiley & Sons, Inc., Hoboken, New Jersey, 2013

IT Environments for BIM in FM

Markus Krämer, Thomas Bender, Nancy Bock,
Michael Härtig, Erik Jaspers, Stefan Koch, Marko Opić
and Maik Schlundt

4.1 The Digital Twin

The use of BIM models for the management of buildings throughout their lifecycle is often considered in the context of so-called Digital Twins (see Sect. 3.3.6).

The principle of a digital twin was first defined by Michael Grieves in 2002 in the context of the inauguration of a Product Lifecycle Management (PLM) Center at the University of Michigan, although the term digital twin was not used at that time (Grieves and Vickers 2017). The prerequisite for the concept was that every system consisted of two sub-systems—the physical system that has always existed, and a new virtual system that contained all information about the physical system.

M. Krämer (✉)
Hochschule für Technik und Wirtschaft Berlin, Berlin, Germany
e-mail: markus.kraemer@htw-berlin.de

T. Bender
pit – cup GmbH, Heidelberg, Germany
e-mail: thomas.bender@pit.de

N. Bock
BuildingMinds GmbH, Berlin, Germany
e-mail: nancy.bock@buildingminds.com

M. Härtig
N+P Informationssysteme GmbH, Meerane, Germany
e-mail: michael.haertig@nupis.de

E. Jaspers
Planon B.V., Nijmegen, Niederlande
e-mail: erik.jaspers@planonsoftware.com

M. May et al. (eds.), *BIM in Real Estate Operations*,
https://doi.org/10.1007/978-3-658-40830-5_4

The digital twin is the virtual representation of a physical object or system during its lifecycle (planning, construction and operation) using real-time operational data and other sources, in order to enable understanding, learning, inferring and dynamic recalibration for improved decision making (Mikell 2017). The physical object can be anything—from a building to a ball bearing, while the system can be, for example, electrical, mechanical or software-based. Furthermore, the interoperability between these systems (i.e. systems of systems) is taken into account.

The following aspects can be derived from the above-mentioned definition as essential components of a digital twin:

- Digital/virtual representation (e.g. 3D BIM model, enriched with semantics) and the bidirectional exchange of information with the physical world,
- Consideration over the entire lifecycle of the object/building (planning, construction, operation),
- Consideration of a temporal component (real-time capability for data acquisition, analysis and processing).

The digital twin technology has many applications not only in construction and FM (cf. NN 2021j). It is important to mention that the digital twin technology for buildings does not necessarily imply the use of BIM models. There are different types of implementations for digital twins around buildings. They can be mathematical-numerical models or even based on simple 2D drawings of buildings (e.g. floor plans). In addition, technologies such as laser scanning and photogrammetry can be used to create digital representations of buildings (especially existing buildings) and as such are also referred to as a digital twin.

When deciding which type of digital twin to use, it is important to determine the purpose of the digital twin beforehand. If, for example, the purpose is to predict a failure due to wear and tear during the operation of a pump, an IoT-based numerical model could work well.

For example, when operating a power plant, the applied digital twin technology is a combination of plant operation data that match the logical (SCADA—supervisory control and data acquisition) representation of the system. However, one would not use a

S. Koch
Axentris Informationssysteme GmbH, Berlin, Germany
e-mail: skoch@axentris.de

M. Opić
Alpha IC GmbH, Nürnberg, Germany
e-mail: m.opic@alpha-ic.com

M. Schlundt
DKB Service GmbH, Berlin, Germany
e-mail: maik.schlundt@dkb-service.de

BIM model to operate the power plant. Complex digital twins can be established by integrating BIM models (cf. Fig. 4.6). Digital twins offer various new potential benefits.

4.1.1 Representation of Physical Components in the Virtual Model

Sect. 4.3 discusses methods for providing BIM models in a common data environment for the lifecycle of buildings. In fact, the transformation of BIM models, starting from numerous discipline and coordination models (see Sect. 3.4.5) is a central step in creating a building digital twin to represent physical building components in a virtual model. Here, the components used in a CAFM system for operations are also of great importance. The benefit of the digital twin with regard to the mapping of physical structural geometry, for example, is to give access to the real environment via a virtual environment for indoor navigation and training purposes. In connection with Augmented Reality technologies (see Sect. 2.6) it is also possible to provide *remote assistance* for on-site personnel (see Sects. 6.3.1 and 10.2.4).

4.1.2 Tracking and Analysis of Component Behavior Through IoT

In the context of the digital twins, there are two types of data for describing assets and related objects: business- and process-related data, as they are typically managed in CAFM systems (e.g. object classifications, model information, and maintenance schedules), as well as the dynamic (IoT) parameters that are captured and processed in real time. These include, for example, temperature, humidity, electrical power, CO_2, and vibration values. The analytical evaluation of these variables can be used to detect events that cause follow-up actions in the CAFM system (see Sect. 4.1.4).

4.1.3 Monitoring and Analysis of Building Parts Through Linking IoT Data with BIM Models

The actual behavior of objects, as far as it is known, can be mapped in the model. An example of this was shown in a GEFMA Future Lab (May and Turianskyj 2017) (see Fig. 4.1). The temperature and the CO_2 concentration were recorded live in a conference room and linked to the digital twin.

For this, the IoT data must be assigned to BIM objects in the model, such as the temperature of the bearings of a pump to the corresponding BIM object. This link can be established directly between the IoT platform and the BIM object. An alternative is to use the asset components in a CAFM system for this purpose. Various CAFM systems already offer corresponding IoT functions today. If the BIM models are already connected to the CAFM/IWMS system, the IoT data can be linked without a direct con-

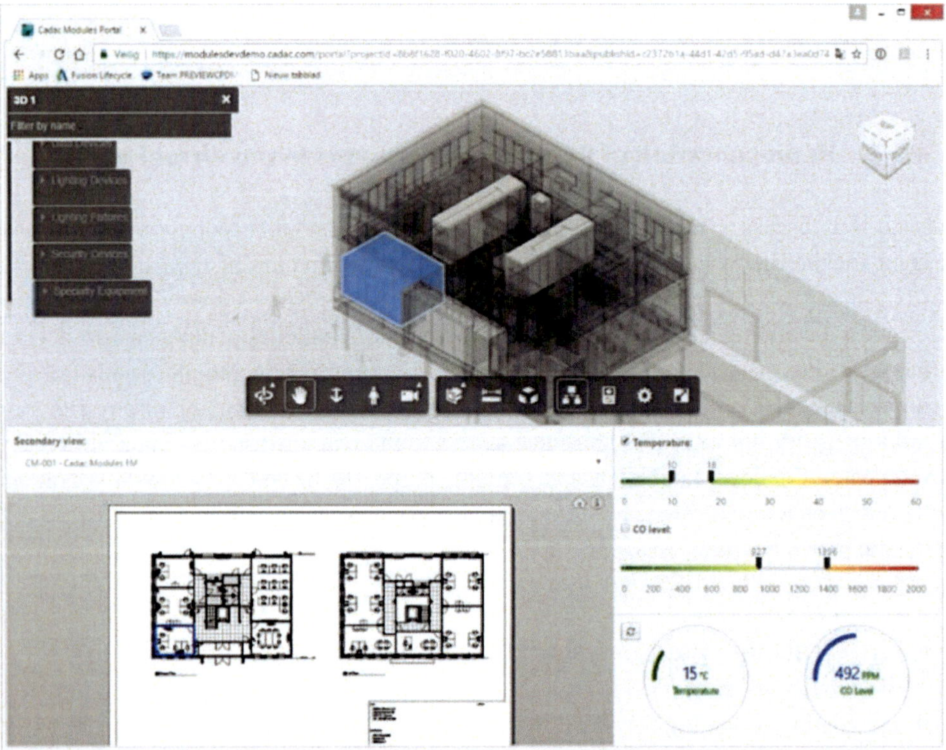

Fig. 4.1 IoT data linked to a BIM model enabling behavior analysis

nection between the IoT platform and the BIM model. This approach can prove to be efficient and cost-effective.

4.1.4 Automation of Event-Driven Actions

Different CAFM systems offer IoT services by providing IoT functions on their own platform or by communicating with IoT platforms from third-party providers. Some providers even offer both options at the same time. This enables the evaluation of events whose data is captured by IoT in the CAFM system (cf. Fig. 4.2).

Advanced integrations between IoT platforms and CAFM systems enable the automation of actions in the CAFM system to identified events from the IoT platform. An example of this is starting maintenance tasks for an asset when the transmitted event indicates a possible failure of this asset in the near future.

The combination of this functionality with the ability of locating affected assets on site via a BIM model (cf. Fig. 4.3) provides an effective IT environment for operations.

The BIM method provides numerous starting points to create efficient digital twin models. However, before projects are started in this area, the purpose of using a digital twin should be defined first. Only then necessary transformations of the BIM models can

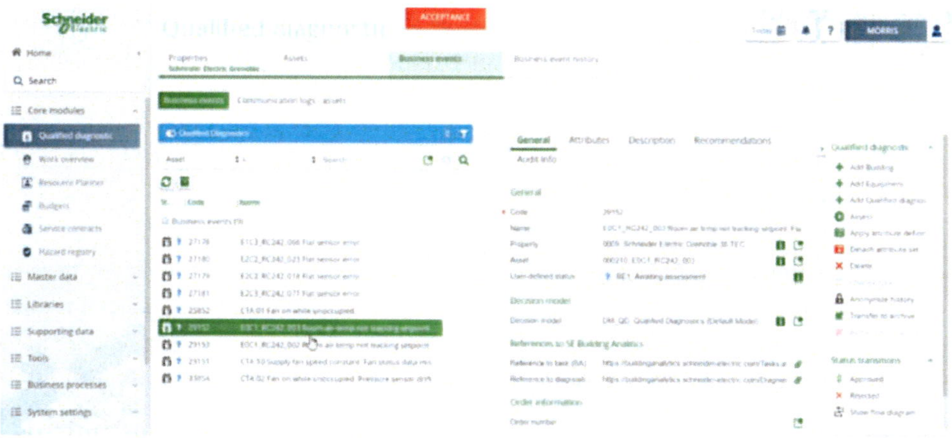

Fig. 4.2 Events from an IoT infrastructure that are processed in the CAFM system

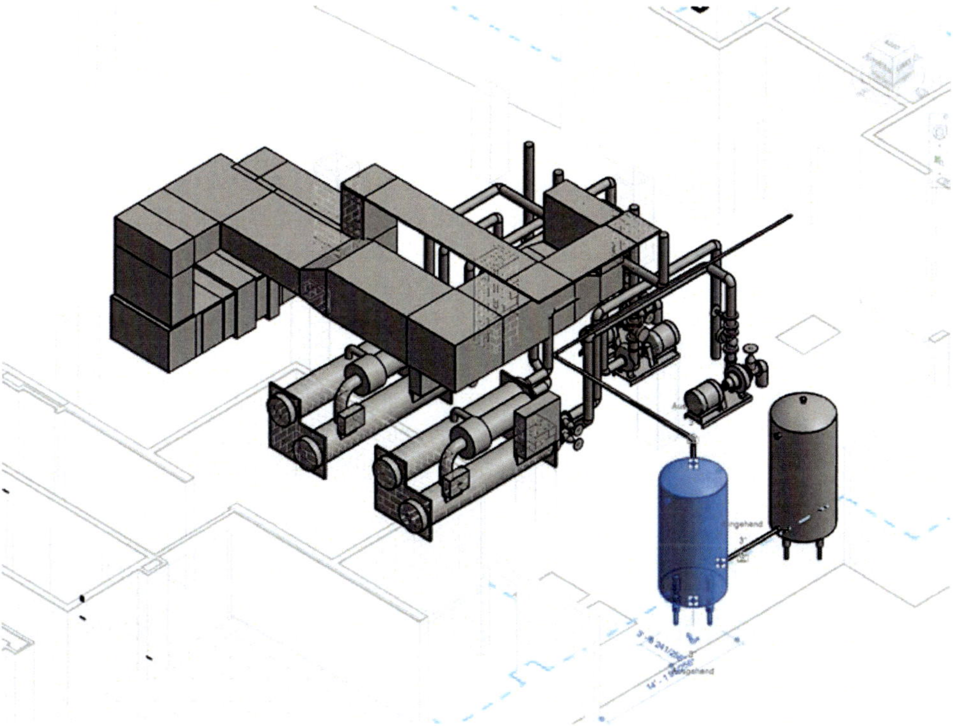

Fig. 4.3 Representation of a distributed technical system in a BIM model

be specified, an appropriate implementation be planned and the IT components required be selected. For this purpose, a business case (see Chap. 6) must be created in which the effort for implementation is set against associated costs.

4.2 BIM Tools

In Sect. 4.1 the enormous potentials of a building digital twin were explained with regard to the operational phase of buildings. An important, but not mandatory, part of a digital twin is the combination with building information models. BIM represents a digital way of working which delivers as a result many digital, 3D BIM models. In a typical project there are various discipline models, e.g. from architechture, engineering, HVAC, etc. In addition to the geometric representation, the building information models also contain the associated alphanumeric object information and semantic relationships of the objects to each other (see Sects. 2.2 and 3.4).

For the creation, management and use of digital building information models, various software tools (BIM tools) are involved in the BIM lifecycle. In the respective use cases resulting tasks and digital deliverables are to be created by specialized tools depending on the BIM processes. For a digital way of working without information loss, it is essential that these systems can interact with each other via standardized interfaces and data formats (e.g. via IFC in an Open-BIM project, see Sects. 3.3.4 and. 5.3.1).

BuildingSMART has developed a software certification procedure to ensure smooth data exchange based on IFC, according to which a large number of software products supporting the BIM method and data exchange with IFC are already certified. An overview of the products for different BIM applications is available from buildingSMART (NN 2021at).

The systems range from BIM authoring tools to calculation and simulation systems. To classify BIM tools, four categories are used (see Fig. 4.4).

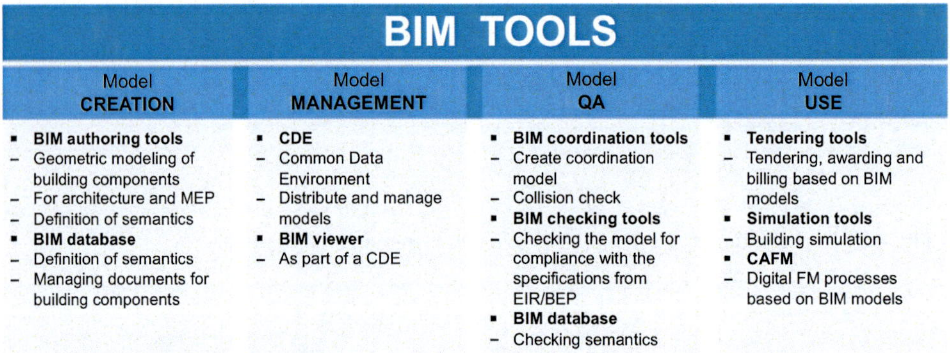

Fig. 4.4 Classification of BIM tools

- Model creation,
- Model usage,
- Model management,
- Quality assurance.

4.2.1 Tools for Model Creation

Since the essential delivery in a BIM project are the building information model(s) with their geometric and alphanumeric information, the tools for their creation are key to success.

4.2.1.1 BIM Modeling Tools for Architecture

These tools are used to create and plan buildings or architectural models with parametric 3D objects. The focus is on geometric modeling of building components. However, authoring tools also have functions for defining semantics (attributing components and relationships between components). To define or capture alphanumeric information, BIM databases specifically set up for this purpose are increasingly being used in BIM projects, which are linked to the BIM modeling tool (see Sect. 4.2.1.4).

Authoring tools store the building information model they create, in this case the discipline model *architecture*, usually in the proprietary, native file format of the software manufacturer. In order to ensure lossless data exchange in an Open-BIM project, it is important that these tools provide a standardized IFC interface for import and export (see Sect. 4.2.3.1).

4.2.1.2 BIM Modeling Tools for Building Technology

These tools are specialized in modeling objects of building technology. As a rule, building information models of the discipline architecture are the basis, which are first imported and then enriched with the required assets. Often, these programs also contain special trade-specific modules for calculating and dimensioning individual technical assets and components. They also support the connection between equipments, e.g. by means of pipes or ducts through assistance systems during modeling. As a result, a building information model of the discipline MEP is generated.

4.2.1.3 BIM Modeling Tools for Structural Design

These tools support the workflow of structural design across various planning phases, such as preliminary, approval or execution planning. A special feature of these modeling tools, in addition to the discipline-specific orientation, is the ability to deal with both geometric and analytical models. Most tools therefore support the import of architectural models and, usually in dialog with the user, transform geometric objects such as walls into corresponding analytical model elements. For example, a 3D wall can be represented as a 2D pane, plate or 3D FEM network element for the corresponding calculation methods (see Sect. 2.11.2.2). A particular challenge is the positioning of analytical elements and their connections, which is required for the calculation but not always clearly defined as a result of the transformation.

4.2.1.4 BIM Database as an Addition to the Authoring Tools for Alphanumeric Object Information

In the BIM database (BIM DB), the alphanumeric object information (semantics) of the building components modeled in the respective authoring tool is recorded and managed. The BIM DB and authoring tool are connected via interfaces, e.g. via a BIM DB plugin installed in the authoring tool.

By separating geometry and semantics into two different systems, the geometric model must be extended only slightly in the authoring tool. In addition, a BIM database offers further functions to easily and efficiently capture, change and evaluate alphanumeric data. In most cases, database systems offer typical evaluation tools such as report generators with which even complex component or quantity evaluations can be created easily. Database systems usually rely on relational (table-oriented) data models and can be queried using a standard query language such as SQL.

For the use of a BIM database in a BIM project, it is important that it has interfaces for bidirectional data exchange with authoring tools. Only in this way can additional alphanumeric attributes from the BIM database be mapped back to the parametric objects in the authoring system.

4.2.1.5 BIM Object Server and BIM Object Libraries

BIM object servers (component servers) provide BIM objects for architecture or MEP. Often manufacturers make their equipment or components available as BIM objects for download, partly in the proprietary formats of specific authoring systems (e.g. Revit). But BIM objects are also offered in the open IFC standard format.

In contrast to the BIM server, BIM objects are only made available to users for download in order to integrate them into their own model project using an authoring tool. For example, manufacturers of walls, ceilings, foundations, roofs or windows offer BIM databases for their products.

4.2.2 Tools for Model Management

When using the BIM method, collaboration of the project participants is mandatory. Model data created by the individual participants, must be checked in terms of quality, combined in coordination models and made available to the participants. To be able to transparently control these complex processes, specially developed so-called collaboration tools are used in the BIM project.

4.2.2.1 Collaboration Software, Project Spaces and BIM Servers

Various tools are used for collaboration, which are usually used in a tool chain. Tools that provide functions for storing, exchanging and sharing BIM models or parts of it for all involved disciplines are often referred to as BIM platforms or project spaces. This also often includes document management.

BIM project spaces usually handle different proprietary data formats, but increasingly also the open IFC format. Other functions of project spaces include permission management for collaboration as well as maintaining the version history of objects and models.

In comparison to most project spaces, BIM servers do not treat the individual BIM discipline models as a file (container) that is managed, but store the model elements themselves in an internal database. In this way, different discipline models (architecture, MEP, …) can be managed in a database and combined with each other at the level of model elements for certain tasks. For example, so-called coordination models can be used to detect collisions between components of different disciplines (discipline models).

Often BIM servers are part of the collaboration software that is used for concurrent, up to simultaneous editing of the building models with different software tools. Changes to models in the BIM server are tracked and users can access the change history.

These tools together provide the Common Data Environment (CDE) for BIM projects which is examined in more detail in Sect. 4.3 with regard to benefits, requirements, tasks and integration with CAFM systems in the operational phase.

4.2.2.2 BIM Viewer

Models that are downloaded from a BIM project space can of course be accessed and edited with the respective authoring tool. However, if the authoring tool is not available or editing is not needed, a BIM viewer is used. BIM viewers are often available free of charge for native BIM formats, but also for IFC.

IFC viewers can read IFC models and show their contents both graphically and alphanumerically. Some of these tools also allow alphanumeric attributes to be changed and data to be exported, e.g. in COBie format (see Sect. 5.3) as well as creating reports and analyses of the BIM model. They often allow the client to "walk through" the building virtually using the model.

Some popular IFC viewers can be extended to a BIM model checker by activating additional, paid-for functions (see Sect. 4.2.3.2), with which collisions or compliance with modeling standards can be checked.

4.2.3 Tools for Quality Assurance of the Models

4.2.3.1 BIM Coordination Tools

The use of various discipline-specific authoring tools in BIM projects and the discipline models created with their help must be coordinated with each other and checked for consistency. This applies to BIM model elements that occur in several discipline models (e.g. walls, ceilings), but which differ in the individual discipline models with regard to their semantic and geometric content (e.g. a wall object in the discipline model architecture and in the discipline model of structural engineering). But this is valid also for

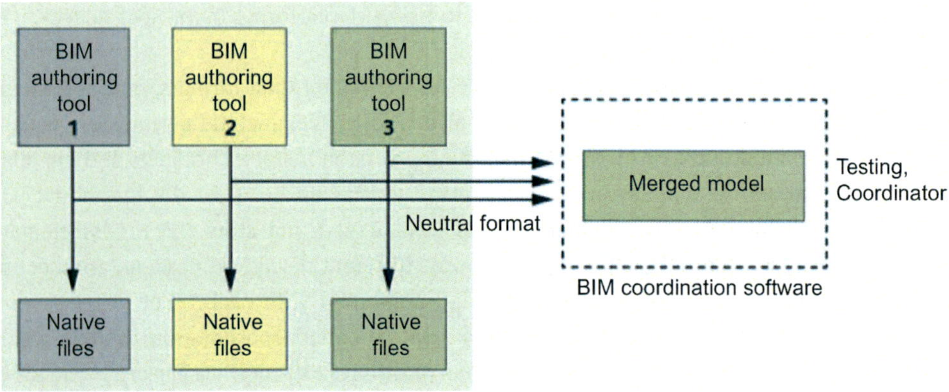

Fig. 4.5 Use of proprietary and vendor-specific data formats in authoring tools

model elements that only occur in one specific discipline model (e.g. a duct or pipe element in the MEP discipline model).

For these reconciliation processes, BIM coordination tools are used. With their help, the different discipline models are integrated in coordination or federated models (cf. Fig. 4.5, NN 2019k). The specialist models, which are usually available in native formats, can be converted into a neutral, open format using the authoring tools themselves via an IFC export, in order to then be merged with the coordination software in the coordination model. However, many coordination tools also have the ability to read specialist models directly in their native format in addition to importing IFC models.

In the coordination model, the elements of different specialist models can be viewed and analyzed together, e.g. collisions can be checked. In this way, collisions are detected early, e.g. missing wall breakthroughs or collisions of model elements, and the corresponding solutions can be found. The open BCF format (BIM Collaboration Format) is often used for communication of the test results (cf. Sect. 5.3.2).

Coordination tools are usually a permanent part of the collaboration software explained in Sect. 4.2.2.1.

4.2.3.2 BIM Model Checker

In addition to coordination tools or as part of such tools, BIM model checkers are used. They enable automatic checking of requirements specified in EIR or BEP for the level of elaboration as well as compliance with the specified modeling guidelines. Depending on the authoring tool used, the test tools can be a part of the authoring tool, but standalone software products are also common. For review of the alphanumeric BIM data, the content of the BIM DB can be checked in some cases as well.

The type and extent of requirements for the building information model vary greatly from project to project depending on the supported BIM use cases, so that BIM model checkers are required to configure predefined test rules flexibly and, if necessary, also define new rules. Some tools have begun to offer not only simple syntactic checks (e.g. the presence of certain attributes), but also complex rules for compliance with relevant standards and guidelines. For example, in the field of FM, compliance with space standards can be automatically checked by rules.

The test results, including individual annotations can then be transferred to other participants via the Open-BIM format BCF. The BCF format does not transport the entire model, but only relevant sections and references to affected model components.

4.2.4 Tools for Model Usage

The models created in a BIM project serve as a data basis and are accessed for further processes and use cases during the planning, construction and operation phases. For these follow-up processes, suitable additional software tools are used. For example, some software tools use the building information model to generate specifications and tender documents based on the already validated data of the models or to prepare a maintenance schedule in a CAFM system. In the following, some typical tools for model usage are explained.

4.2.4.1 BIM-CAFM Software

CAFM systems with a standardized BIM interface are often able to import and export IFC, COBie, and/or CAFM-Connect data. In addition, they usually integrate BIM viewers, with which corresponding models can be viewed. Some CAFM systems also have proprietary interfaces to specific BIM authoring systems or BIM-enabled CAD systems. Thus, changes to the alphanumeric component data can be made in the CAFM system as the leading system during the operational phase and returned to the BIM model if there is a bidirectional interface. If there are major changes to the building, current data from a CAFM system, e.g. as an IFC file, can be exported and made available for planning.

Basic BIM functions of CAFM systems can be certified according to GEFMA guideline 444, catalog A15 (BIM data processing) (NN 2020a). Further information on this procedure is described in Sect. 8.2. Hence, modern CAFM systems already offer functions that include elements of a BIM DB and in part even a BIM server. CAFM systems are therefore also candidates for providing an operational CDE, which is described in Sect. 4.3.2.

4.2.4.2 Project Management

This refers to software for planning and scheduling of construction projects, including cost management. BIM-enabled project management software offers so-called 4D or 5D planning, in which the three geometric dimensions (3D) are extended by the dimensions project planning (4D) and cost planning (5D) (see Sect. 3.4.3). As a rule, such project management tools import a BIM model of an authoring tool to extend it by the aforementioned dimensions.

4.2.4.3 Simulation Tools

These software tools process building information models to enable simulations of certain processes related to the real estate lifecycle. These include, for example, simulations and analyses based on energy consumption, heat demand, acoustics, lighting, fire protection, environment, factory/facility planning, collisions or project progress. Further details on BIM-based simulation tools can be found in Sect. 2.11.

4.2.4.4 BIM Software Toolkits

These are software libraries or plugins that can be integrated into the users' own software. This includes, for example, functions such as visualization, explorer or BIM export/import options, which do not have to be redeveloped.

4.3 Common Data Environment and BIM-CAFM Integration Possibilities

The application of the BIM tools for model creation and editing presented in Sect. 4.2 as well as the model-based information exchange without information loss in BIM projects between these tools require a common data environment. The term "common data environment" (CDE) was first introduced by the British BIM specification in BSI PAS 1192 in 2007 (NN 2014b). Today, the term CDE has been adopted by almost all national and international standards and guidelines: DIN EN ISO 19650-1 Section 1 (NN 2019a), DIN EN ISO 1950-2 Section 5 (NN 2019j), DIN SPEC 91391 (NN 2019i), VDI 2552 Sheet 5 (NN 2018f).

A CDE provides a central, shared data environment for the

- Collection,
- Management and
- Distribution

of model-based information in a BIM project or, particularly relevant for the operational phase, for an asset. At least conceptually, the CDE supports the entire lifecycle of a building.

In order to fulfill these tasks, the previously described BIM platforms or BIM project spaces are used to implement a CDE, with the integration of other BIM tools, such as BIM servers, BIM viewers or BIM test tools, also being common in the CDE (see Sect. 4.2.2). However, in practice, one does not find just one CDE, but rather, parallel or successively, different CDE's from different manufacturers are used in the various project phases. For this reason, not only the data exchange between the various BIM tools and the CDE, but also between different CDE's, must be guaranteed. For this, too, open standards such as IFC, CAFM-Connect, COBie (see Sect. 5.3.3) or OSCRE are used.

In the first service phases of the building lifecycle, during planning and construction, the CDE supports project structures. Consequently, the resulting digital building information models are also referred to as project information models (PIM), which are then managed with the CDE. With the transition to the operational phase (FM handover), however, the focus shifts to the consideration of the building as an asset. For this reason, during the operational phase, it is called asset information model (AIM). This also changes essential requirements for the CDE. In addition to collaboration processes that are repeated more regularly and, in most cases more standardized during the operational phase, the integration of the AIM with dynamic process and consumption data as well as economic information becomes more important. Therefore, Patrick Mc Leamy describes the AIM as the "Building Operation Optimization Model" (BOOM).

Consequently, during operation, the CDE must also include CAFM and ERP systems, which either work together with the CDE or even take over essential tasks of the CDE itself. For the operational phase, this is often referred to as an asset management system (AMS) and sometimes refers to the CDE of the operational phase (Fig. 4.6). The special importance of modern BIM-capable CAFM systems as an important element of the AMS or the CDE in the operational phase is discussed in Sec. 4.3.2 explaining different integration scenarios for combining BIM and CAFM.

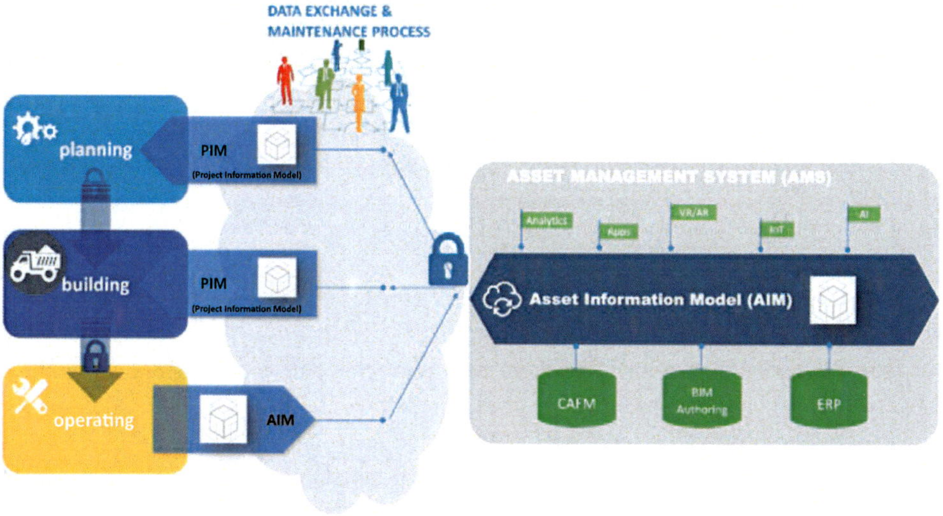

Fig. 4.6 Asset management system (AMS) as a representation of a CDE in operation

4.3.1 Benefits, Tasks and Development Stages of a CDE in the Operational Phase

The use of a CDE in the operational phase is generally expected to result in a significant reduction in time and, consequently, costs, inter alia, through the automation of administrative tasks for information delivery from the perspective of both provision and acceptance (approval) of digital building models. Benefits, such as the avoidance of claims, incorrect or ambiguous object names as well as the reduction of post-processing costs due to information losses as well as can be expected. For this purpose, at least partial automatic checking of incoming building models for correct syntax and semantics (quality assurance) is required. In the operational phase, the CDE is also expected to reduce efforts for processing information requests and to improve the correctness of decision-making through a reliable and up-to-date information base.

To achieve this benefit, the following requirements and tasks arise for a CDE during the operational phase:

- Unambiguity, correctness and integrity of the managed digital as-built building information (single source of truth),
- Traceability and transparency of all information deliveries, including associated responsibilities and copyrights (audit trail) and a suitable management of access rights,
- Revision-proof and legally ensured documentation of critical information that is required, for example, in the context of the operator's responsibility,
- Support of quality assurance and model checking of deliveries of new or adapted model content (e.g. with regard to classification, completeness, correctness, and timeliness).

For this purpose, the CDE maps workflows that support collaborative modeling according to DIN EN ISO 19650-1: 2019 over four defined statuses

- Shared,
- In progress,
- Released and
- Archived.

The principle workflow is illustrated in Fig. 4.7.

A CDE uses information containers as a core element to meet these requirements and map collaborative workflows. Such information containers include defined, content-related sets of information (container content) and metadata that classify the information content and, for example, provide information on the status of discipline models in the BIM process. Additional metadata may include responsibilities and any other properties (see Fig. 4.8, based on NN (2019i)). Information containers of a CDE are considered the smallest unit of information that can be individually retrieved.

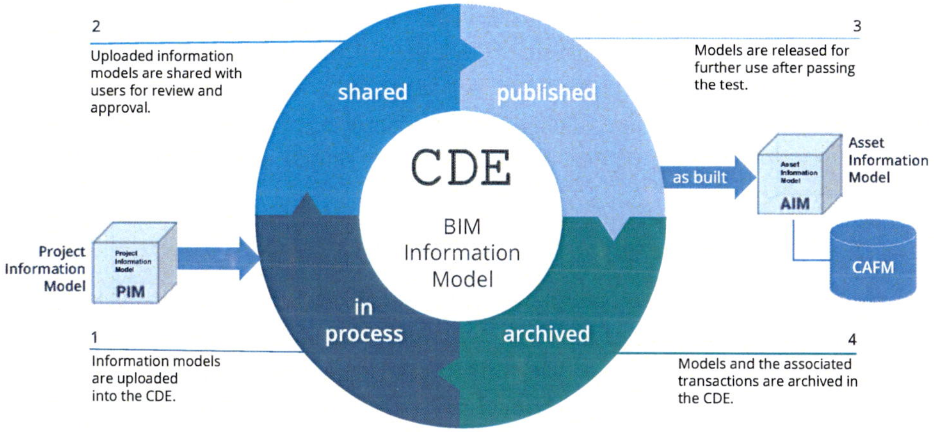

Fig. 4.7 Status of collaborative modeling in a CDE

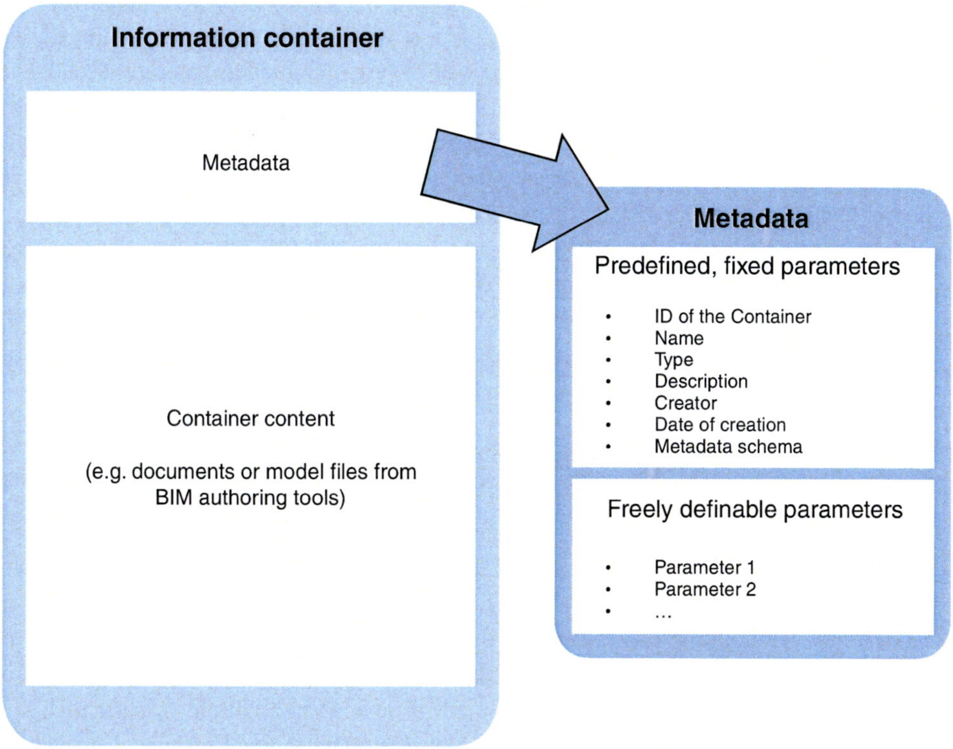

Fig. 4.8 Principle of an information container

Analogous to the three development stages of information management within the BIM method (see NN 2019j and Sect. 4.2) or the BIM maturity levels according to Bew and Richards (see NN 2014b and Sect. 3.4.2), different development stages of a CDE are also distinguished, with Stage 2 being container-based and Stage 3 being database-based. Stage 1 is not separately defined because no model-based information exchange takes place at this stage.

4.3.1.1 Container-Based CDE (Level 2)

In the simplest case, entire building models are managed as files and documents in information containers at this level. The information container can include one or more files/documents, for example in PDF format (drawings, technical data sheets, etc.). Furthermore, CAD models in DWG format (2D, 3D) or entire parametric BIM models in the proprietary format of the BIM authoring system (e.g. RVT in the case of AutoDesk Revit) are common.

In this case, the search for information and model elements can only be carried out using the metadata of the information container. The content of the model (e.g. individual walls and rooms) is not available for a search or filter operation in the CDE at this level. Linking several discipline models of different containers can only be achieved based on the metadata of the containers. At this level, IFC models can already be managed (openBIM), but in case of a CDE level 2 possibly included links within IFC models are unknown could not be used, even when contained in the IFC models (e.g. via the globally unique identifier (GUID) of building elements).

In the simplest case, a level 2 CDE can be provided by a document management system (DMS), with documents being files of BIM discipline models. BIM project spaces and BIM platforms usually master at least level 2, but in comparison to a simple DMS, BIM project spaces have specific BIM related functionality supporting processes as well as specific BIM metadata for mapping workflows.

4.3.1.2 Database-Based CDE (Level 3)

With level 3, the information containers are dissolved and individual building objects from the models are stored as entities in a database. In this way, individual BIM objects of a model (e.g. wall, room, asset) can be individually addressed in the CDE via the database. With the dissolution of the information containers, it is now possible to search, filter or reference building components in the CDE. In order to achieve this, BIM servers and BIM databases are used in the CDE (cf. Sect. 4.2.1.4). A typical CDE of this level also masters the container-based management of level 2 in addition. In particular, vendors of comprehensive BIM toolkits usually offer a CDE solution of level 3, sometimes limited to the proprietary data formats of their own BIM authoring tools (closedBIM). However, a trend towards the management of models in the open IFC format can also be observed (openBIM).

4.3.2 BIM-CAFM Integration to Establish a CDE for the Operational Phase

The architecture, engineering and construction industry is increasingly relying on BIM-based planning and design tools today. As a result, there are usually various BIM discipline models of buildings available that facility managers can use throughout the building's lifecycle. Implementing a BIM strategy for managing buildings throughout their lifecycle requires careful planning. It is very important to define the main goals for the implementation of BIM, as this defines to a large extent the type of integration, the toolsets required, and the costs associated with the project.

In general, the transfer of BIM models to CAFM systems requires, in addition to the transfer of data, also geometric adaptation of the model (cf. Fig. 4.9), in order to be able to use it throughout the lifecycle.

In addition to BIM platforms and BIM project spaces, which are often used as a CDE for the planning and construction phases, BIM-capable CAFM systems and, to some extent, ERP systems already offer numerous functions that support or even completely map specific requirements of the operational phase. In the following, three typical scenarios will be presented how BIM-capable CAFM systems can be used for a CDE in the operational phase (Aengenvoort and Krämer 2021). In all three scenarios, the CAFM database is used to map the non-geometric, alphanumeric information of the digital building models (AIM) at the level of individual building objects (e.g. rooms, assets), which corresponds to level 3 of a CDE. CAFM systems already map the building topology (property, building, floor, room, etc.) and numerous, flexibly extendable object types, e.g. in the field of asset engineering, in their database. In addition, CAFM systems

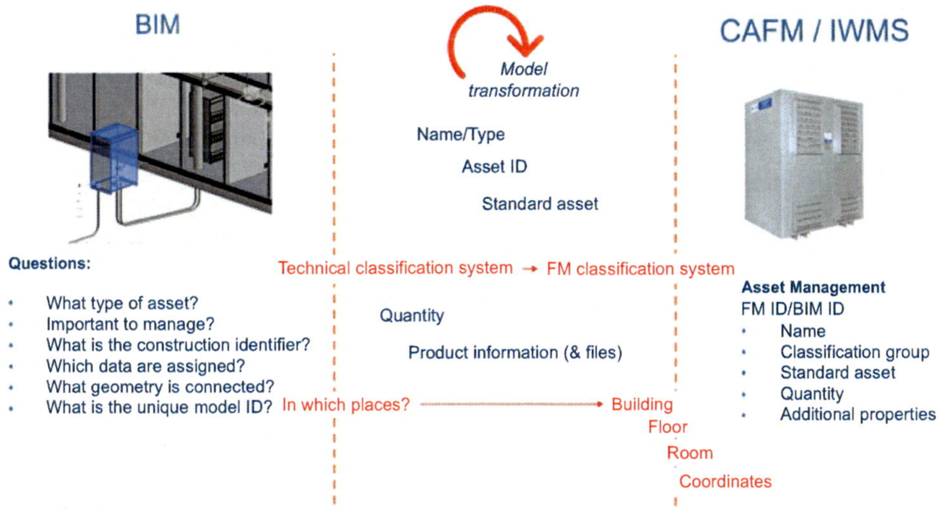

Fig. 4.9 Adaptation of the BIM model for use in CAFM systems

enable to assign documents and CAD drawing files to database objects, thus providing container-based CDE functions of level 2.

4.3.2.1 Integration Scenario 1: BIM-CAFM Handover by Information Extraction (FM Handover)

In this scenario, the BIM model is used to extract as many building data as possible so that they can be imported into the CAFM system (cf. Fig. 4.10). This may also include existing floor plans. For geometric information from an as-built PIM, 2D/3D CAD plans may be generated once and then maintained over the lifecycle if necessary. After extraction, the model is archived but no longer used for interaction. It can of course be called up at any time and, for example, viewed using a BIM viewer. In most cases, this is the approach with lowest costs and lowest integration complexity.

CAFM systems that follow this integration approach aim for the most efficient execution of FM handover at the end of the construction phase. Therefore, the relevant structured, alphanumeric information from the PIM is transferred directly into the CAFM database by using either services from a BIM server platform or specific interfaces from CAFM providers. This not only applies to asset data, but can also include spatial data such as the areas of rooms. With COBie (see Sect. 5.3 and NN 2021e) there is an internationally accepted structure for exchanging alphanumeric BIM data. Many CAFM providers support BIM data exchange via the COBie format or the COBieLite format. In Germany and the DACH region, the CAFM-Connect format is also increasingly gaining acceptance (see Sect. 5.3 and NN 2021f). In addition, software vendors may offer alternative implementations. It is important to ask which formats of the BIM authoring tools are supported by the CAFM provider and to ensure that they match the actually available model formats.

In *Integration scenario 1* there are therefore two separate models during the lifecycle: the native, initially unmodified as-built PIM, which may have to be updated manually in

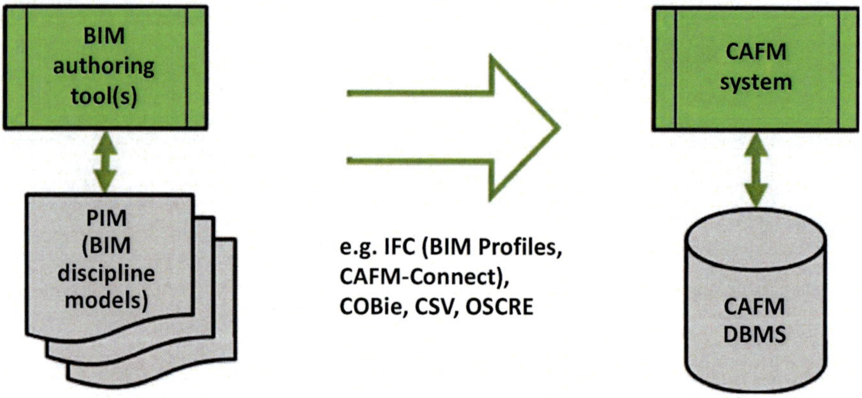

Fig. 4.10 Integration scenario 1: BIM-CAFM handover "FM handover"

parallel with the respective BIM authoring software, and an AIM, which is mapped into the CAFM database and linked with CAD documents.

4.3.2.2 Integration Scenario 2a: Use of the Source Model (native) Over the Life Cycle—Partial CAFM-BIM Integration

In this approach, the source models are used in their proprietary formats of the BIM authoring tool. In this case, probably some data extraction is also required to fill the CAFM database, but the BIM model (PIM) itself is linked to the CAFM system. So modeling software such as Autodesk Revit, Graphisoft ArchiCAD or Nemetschek Allplan is typically supported. For the implementation of proprietary interfaces, plug-ins of the CAFM software vendor are integrated into the respective BIM authoring tools, with the help of which individual or multiple BIM objects can be synchronized with corresponding objects (entities) of the CAFM database (cf. Fig. 4.11).

In most cases, this interface is bidirectionally, so that changes to alphanumeric information in the CAFM database can also be transferred back to the PIM managed with the BIM authoring tool. However, this is not always the case and a point that requires checking. This type of interface design already allows an AIM in the native format of the BIM authoring tool, which can at least be maintained in parts during the operational phase. Alphanumeric information is usually changed via the CAFM database. However, this scenario requires special knowledge of the BIM authoring tool by the employees of the FM department. The scenario thus includes approaches to a CDE of level 2 and, via the CAFM database, partly also to level 3.

For example, if the PIM contains asset data (supplier, type, etc.) and those assets are replaced, it has to be clear who makes use of the data and who is responsible for updating. As explained, this can be done in the PIM with the help of the BIM authoring tool or in the CAFM system or maybe in both systems. In practice this is usually done by a specialist BIM expert.

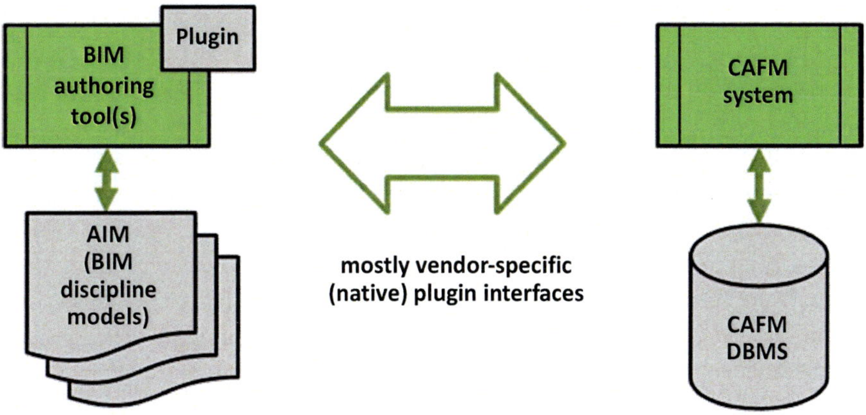

Fig. 4.11 Integration scenario 2: Bidirectional synchronization of BIM models (AIM) with the CAFM database

Another important aspect is change management for the building. For example, if a floor is restructured or redesigned, these changes must be done in the BIM model, otherwise it can no longer be used as a reliable source of information (single source of truth) for the building. Consequently, the model must often be updated according to the latest versions of the BIM authoring tool and the corresponding data formats of the model. This is a considerable effort, as the vendors of the tools release new versions of their software on a regular basis (usually annually, sometimes more frequently).

The second integration scenario allows visualization and localization of objects in the AIM and can be used to dynamically display linked information in the geometry model, such as maintenance information, technical documents, or maintenance schedules. The dynamic color coding of BIM objects based on process information or other linked data sets is also possible.

4.3.2.3 Integration Scenario 2b: Use of the IFC Source Model Over the Lifecycle—Partial CAFM-BIM Integration

The use of IFC is considered when buildings have been designed with different BIM authoring tools. This is not an exception: BIM authoring tools are usually specialized in certain aspects of building design. If all these models are imported into IFC, an integrated (or federated) BIM model is created. All aspects that were discussed when using the source model over the lifecycle (see Sect. 4.3.2.2) also apply here.

There is one aspect in this approach that must be emphasized—the change management of the model. According to the idea of an open standard, IFC models cannot be changed (edited) directly. So if a geometric change of the building is pending, it has to go through the original (native) models, which then have to be converted back into IFC. This can lead to additional complications when integrating into the CAFM system, as the BIM object identifier of the geometric object in the model may change in the worst case. The BIM model identifier of a pump, for example, could have changed in an updated model version, even though the pump itself has not changed. Of course there are ways to deal with it, but this is associated with investments and additional complexity.

In situations where a BIM partial-model has been created using different authoring tools, converting it into the IFC format is the most practical way to use these models during the lifecycle. Thus, all models linked or imported to the CAFM system can be combined and displayed or interact with each other. A prerequisite for this is that the CAFM system supports a viewer technology that allows displaying IFC models.

However, this scenario is complex in terms of change management. If the building or a floor is being reconstructed, the associated models must be changed using the native models in the format provided by the authoring tools. The reason for this is that IFC modeling in this case was designed for collision detection during the design phase. The different models have to be combined and checked for dimensional accuracy. Therefore, IFC models can generally not be changed easily with regard to their geometry.

After the change, the sub-models have to be converted back into IFC and then the transfer into the CAFM system has to be carried out again. The combination of data and their representation in one model provides a good understanding of the actual situation of the building (cf. Fig. 4.12).

In addition to data transfer, as explained in Sect. 4.3.2.1, the model has to be linked to the building, asset and room data of the CAFM system. This requires the transfer of BIM object identifiers to the CAFM system to enable the extraction of the model from the CAFM system. This represents an additional step in the data transfer from the model. In addition, the CAFM system must support a viewer technology that can access the model, in this case a model with the proprietary format of the BIM tool.

4.3.2.4 Integration Scenario 3: Use of the Source Model Over the Lifecycle—Collaboration Platform During Operations

The third integration scenario makes the CAFM system an integral part of an asset management system (cf. Fig. 4.13). In this integration approach, a BIM model server is usually used, in which the BIM models of the operational phase are provided as AIM in a database management system (DBMS). Hence, building objects exist parallel in the database of the BIM server and the CAFM database.

Often the BIM models are already stored in this model server in an open standard format, for example the IFC format (openBIM). The linking of elements of the BIM models with objects from the CAFM system, such as maintenance dates, inspection orders or

Fig. 4.12 BIM models provide important insights into the structure and location of (distributed) systems

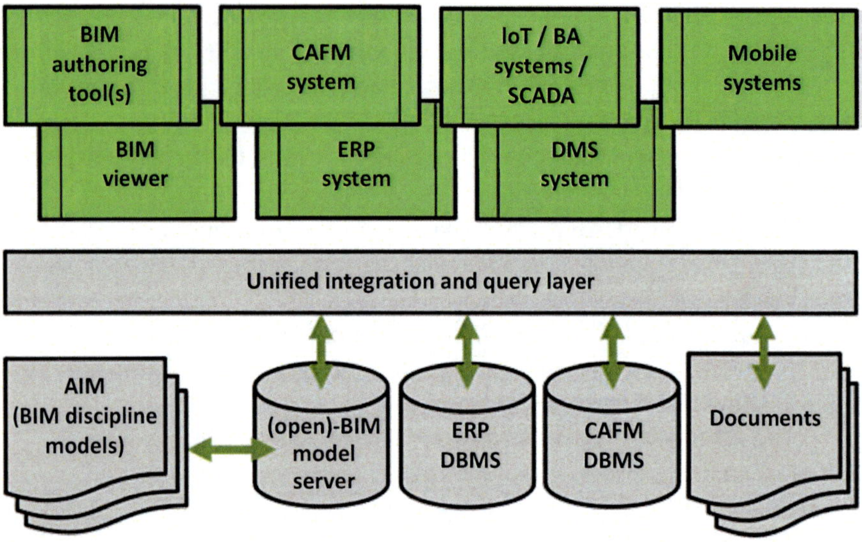

Fig. 4.13 Integrated, distributed (linked data) asset management system

spare parts lists, takes place directly at the level of the participating databases. Analogously, this is also possible for objects managed by an ERP system, in which, for example, contracts, orders or order lists are mapped.

The special feature of this server-based approach lies in the possibility of linking all the required information in the background. The user can access the respective original information sources transparently. For the user, the information retrieval is always the same, regardless of whether the information is retrieved from the CAFM, ERP database or a BIM model server. For this reason, the approach is referred to as distributed data storage according to the "linked-data" principle (Krämer et al. 2018). However, a software must be used as a middleware layer (data integration layer) in order to mediate between the application systems, e.g. CAFM or mobile apps and the information sources (cf. Fig. 4.13).

The CDE initiatives started by BIM software providers aim to provide BIM as a service in which the model interactions, model updates and changes to the alphanumeric data and the associated files of the models are accessible by application programming interfaces (API's).

This *Integration scenario 3* opens up far-reaching possibilities for mapping dynamically configurable workflows, e.g. as part of troubleshooting, processing of service tickets or room bookings by building users. In addition, the integration of BIM models with building automation (BA) systems or with IoT platforms can also be implemented significantly easier and more comprehensive, although the required technologies are not yet fully developed. One problem here may be that each BIM software provider develops their own CDE. A major challenge today—both for CAFM users and providers—is the question of how a structural connectivity between different CDE services can be established. Sect. 8.4 provides an outlook on future initiatives for lifecycle and cross-company collaboration platforms.

4.4 BIM with Free Software

One reason why BIM has not spread as quickly as was first envisaged is perhaps that smaller engineering offices often cannot or do not want to afford the costs of needed BIM software. As an alternative, non-commercial software solutions have established themselves in some areas. Meanwhile, there are also free software for BIM. These should be checked to what extent they can be applied for the respective BIM use cases (NN 2021ay).

The distinction between commercial and free software depends on the licensing model. This regulates the utilization of the software by the user. With non-commercial software, the impression quickly arises that the license costs for a software can be saved and thus the costs can be reduced overall. This should be checked carefully. The following will highlight the advantages and disadvantages and give an overview of what non-commercial software can be used for in terms of BIM.

4.4.1 Open-Source and Free Software

For non-commercial software, there are different licensing models. Every author of software can determine this through the end-user license agreement (EULA). This can contain different conditions, e.g. that the software is only free for private users, restrictions on resale up to the provision of the source code. With open-source software the source code is freely available and can be changed by any software developer as required, depending on the licensing model. The advantage of open-source software is that possible further development of the software by different software developers is desired and possible worldwide, as is the case with the Linux operating system.

Often, programs that are available free of charge are referred to as freeware. However, there is no uniform definition. The license agreement of the software regulates the conditions. Most BIM viewers are, for example, offered as free software, although the source code is not always available. In addition, it must be checked whether the software can also be used free of charge in a commercial environment.

4.4.2 Advantages and Disadvantages of Free Software

At first glance, it seems that software costs, which amount to several thousand euros in the BIM sector, can be saved. There are indeed prominent examples of non-commercial software such as the Linux operating system, which was and still is developed as a non-commercial alternative to Microsoft Windows. In the server environment, Linux is indeed used in commercial environments in practice, on the desktop mostly only by a few private users. The share of Linux is 2.38% in May 2019 (NN 2021au).

In addition, the software costs often do not only involve one-time investments, but also about 10–30% maintenance costs for support services. Alternatively, there are also contracts under which the software is rented and may not be used after the subscription expires. This offers software vendors regular income and thus secures further development.

With non-commercial software, further deveolpments are on a voluntary basis by persons who continue developing the product further or at least finance themselves partly through commercial support services. This is the case, for example, with some Linux distributions.

Another disadvantage of free software is that it often deviates from established user styleguides and interfaces. Depending on the complexity of software, this can make usability more difficult for those used to commercial software. Furthermore, it should be noted whether sufficient documentation of the software exists, the software is used sufficiently widely and thus also enough users contribute to improving the software. Often there are also no training companies, although available video tutorials can partly compensate for this. The mentioned aspects of free, non-commercial software often lead to unpredictable challenges when used in practice. Table 4.1 summarizes the main advantages and disadvantages.

4.4.3 Use of Free Software

The use is worthwhile in organizations that do not have a sufficient budget for software, but do have human and time resources available. In the field of research and development, for example in universities, this can be a decisive argument. By looking at the source code of the software, important concepts can be conveyed or even the further development of the software can be advanced.

Table 4.1 Advantages and disadvantages of free software

Advantages	Disadvantages
• no ongoing software costs • no license costs • often special functions • open-source software is particularly suitable for testing, developing and understanding • use of open standards simplifies data exchange • software can be adapted if the appropriate know-how is available	• often no training provided • only sparse documentation • high learning effort, as divergent from standards • no guarantee of further development • often dependent on a (few) developer(s) • operation (user interface) of the software often does not meet established standards • rudimentary instructions and manuals • learning by doing

Architecture firms that have sufficient IT-affine employees and value individual development can also benefit from the open source community. There is the possibility of adapting the software to one's own projects and acting independently of a software provider.

Software that supports BIM often originates from established CAD programs. In this field there are several free CAD software solutions, for example freecad or LibreCAD. By integrating the open and free of charge IFC format developed by buildingSMART (see Sect. 5.3.1), these CAD-based software solutions can be further developed for the use of the BIM method. Some already support IFC import/export.

4.4.4 Example of 3D Modeling with Blender

From the category *3D modeling*, the free software solution *Blender* is presented in more detail as an example for the use of the BIM method. *Blender* supports the import and export of IFC models through add-ons. In the software, it is even possible to edit IFC objects and attributes, which is enabled by commercial BIM authoring tools only to a limited extent. The software was developed for 3D modeling and animation and offers a large range of functions for this. Meanwhile, Blender is also used in commercial 3D modeling projects, from creation, animation to video editing. The software can be downloaded from the Internet (NN 2021av) and extended to a BIM authoring tool by the IFC add-on.

After installing the add-on, the desired functional extension is available under "File/Import/Industry Found Classes". In Fig. 4.14 a simple model was created in Blender and exported and re-imported as an IFC file. In this way, Blender is suitable for openBIM use.

In Blender, the BIM objects of the corresponding IFC classes of an IFC file can be used and edited directly. It is also possible to link any 3D objects with IFC classes afterwards and adjust the corresponding attributes. Due to the open structure of the plugin, Blender can also be used to analyze the IFC file structure. For example, when the imported IFC file is directly in the source code (see Fig. 4.15).

In comparison with the commercial BIM authoring tool Revit, functions are already provided for large areas of the IFC schema. According to the author, Blender can use about 74% of the IFC schema, while in REVIT only 38% (NN 2021aw). However, the development of the software depends on a single person and the existing documentation must be considered rudimentary. Introduction to Blender takes place via " trial and error". Blender is very well suited to understand the structure and processing of IFC models due to viewing the source code and its further development and offers great potential, especially for research and education.

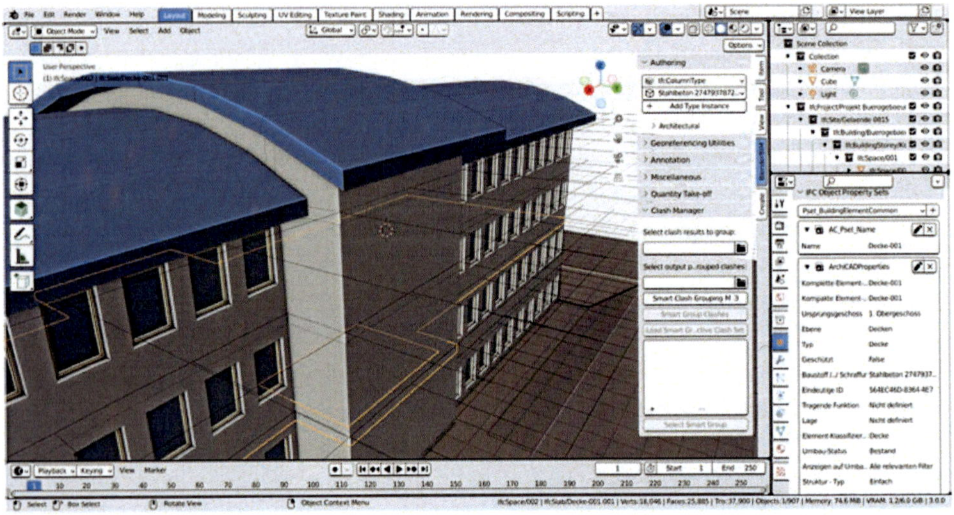

Fig. 4.14 IFC file, loaded into Blender

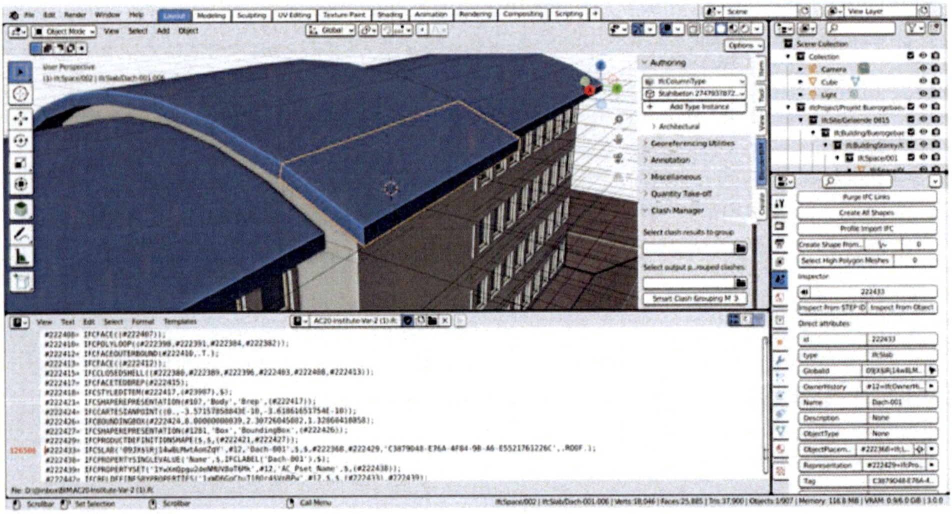

Fig. 4.15 Analyzing an IFC file in Blender

4.4.5 Conclusion

Due to numerous initiatives in the field of openBIM, a manageable amount of free BIM software has already been created. The possible savings in the area of software costs are offset by a steep learning curve at least for the more comprehensive tools. The use of various open-source BIM viewers, which often have a limited range of functions, is also recommended in commercial environments. The further development of open source software solutions is only reasonable if free developer resources are available and individual BIM solutions are needed. In the field of research and teaching, it is definitely an interesting alternative.

4.5 Summary

The use of the BIM method as a driver of digitalization for the operation of real estate and facility management requires, in addition to the introduction of BIM-based processes and corresponding adaptation of organizational structures, the implementation of a suitable IT environment. This chapter deals with the most important aspects required for this.

The starting point is the concept of digital twins, in which models are used as a virtual counterpart to the built, physical environment. Sect. 4.1 focuses on the interaction of technologies and systems from the field of BIM, IoT, SCADA and CAFM. The application of digital twins is explained in analyzing the behavior of buildings and their components, in an extended behavior analysis by IoT up to the automatic reactions by CAFM linking.

In order to answer the question of which software tools could be used for the application of BIM in construction projects and the subsequent phase of real estate and FM operations, important BIM tools are then presented within the phases of model creation, model management, quality assurance and model usage, with the help of which digital tool chains can be implemented.

BIM authoring tools are used for model creation and differ depending on the disciplines considered and their specific discipline models. It should be noted that, in addition to the well-known BIM authoring tools for architectural models (e.g. Revit, ArchiCAD), BIM authoring systems for engineering disciplines such as MEP or structural engineering are expected to map both geometric and analytical or schematic models.

It is obvious that the alignment and coordination of these discipline models are crucial to the success of the project. This is where BIM tools for coordination and model management such as project rooms, BIM servers and BIM platforms come in, which are supplemented by BIM tools for quality assurance. Prominent representatives of this software category are coordination software systems, with the help of which coordination models for clash detection are created, as well as test tools that syntactically and to some extent semantically monitor the quality of the model. Further software tools for model usage in project management, simulation, but also CAFM conclude the presentation of the BIM tools.

All of these tools must interact with each other in BIM projects, but especially in the phase of real estate and FM operations, with as little information loss as possible. For this purpose, a common data environment (CDE) is essential, the benefits of which and development stages are explained. While in the planning and construction phase, BIM project rooms and BIM servers determine the implementation of a CDE in practice, the phase of real estate and FM operations requires the integration of CAFM and ERP systems with BIM. For this purpose, three integration scenarios are presented, starting from a simple FM handover (BIM to CAFM) to an integrated asset information management system as a collaboration platform for operations.

Finally, approaches for the use of free, non-commercial software for BIM in FM are discussed and their potentials and limitations are described.

References

Aengenvoort K, Krämer M (2021) BIM im Betrieb von Bauwerken. In: Borrmann A, König M, Koch C, Beetz J (Eds.): Building Information Modeling – Technologische Grundlagen und industrielle Praxis, Springer Vieweg, Wiesbaden, pp 611–644

Grieves M, Vickers J (2017) Digital Twin: Mitigating Unpredictable, Undesirable Emergent Behavior in Complex Systems. In: Kahlen F-J, Flumerfelt S, Alves A (Eds.) Transdisciplinary Perspectives on Complex Systems, Springer, Cham, pp 85–113

Krämer M, Besenyöi Z, Sauer P, Herrmann F (2018) Common Data Environment für BIM in der Betriebsphase – Ansatzpunkte zur Nutzung virtuell verteilter Datenhaltung. In: Bernhold T, May M, Mehlis J: Handbuch Facility Management, ecomed SICHERHEIT Verlag, Heidelberg, München, Landsberg, Frechen, Hamburg, pp 1–32

May M, Turianskyj N (2017) The Future is Now – CAFM Future Lab 2017. Der Facility Manager 24(Mai 2017)5, 20–23

Mikell M (2017) Immersive analytics: the reality of IoT and digital twin. IBM Business Operations Blog https://www.ibm.com/blogs/internet-of-things/immersive-analytics-digital-twin/ July 13, 2017 (retrieved: 25.06.2021)

NN (2014b) PAS 1192-2 (2014) Specification for information management for the capital/delivery phase of construction projects using building information modelling. London: British Standards Institution

NN (2018f) VDI 2552 Blatt 5. Building information modeling – Datenmanagement. Düsseldorf: Beuth, 22 S

NN (2019a) DIN EN ISO 19650-1. Organisation und Digitalisierung von Informationen zu Bauwerken und Ingenieurleistungen, einschließlich Bauwerksinformationsmodellierung (BIM) – Informationsmanagement mit BIM – Teil 1: Begriffe und Grundsätze, Deutsches Institut für Normung, 2019-08, 49 S

NN (2019i) DIN SPEC 91391-1, Gemeinsame Datenumgebungen (CDE) für BIM-Projekte – Funktionen und offener Datenaustausch zwischen Plattformen unterschiedlicher Hersteller. Deutsches Institut für Normung, 2019-04, 45 S

NN (2019j) DIN EN ISO 19650-2. Organisation und Digitalisierung von Informationen zu Bauwerken und Ingenieurleistungen, einschließlich Bauwerksinformationsmodellierung (BIM) – Informationsmanagement mit BIM – Teil 2: Planungs-, Bau- und Inbetriebnahmephase, Deutsches Institut für Normung, 2019-08, 42 S

NN (2019k) BIM4Infra2020, Teil 10 Technologien im BIM-Umfeld. Publikationen

NN (2020a) GEFMA Richtlinie 444: Zertifizierung von CAFM-Softwareprodukten. Februar 2020, 21 S

NN (2021e) https://www.wbdg.org/bim/cobie/ (retrieved: 27.05.2021)

NN (2021f) https://www.cafm-connect.org/ (retrieved: 27.05.2021)

NN (2021j) https://softengi.com/blog/use-cases-and-applications-of-digital-twin/ (retrieved: 25.06.2021)

NN (2021at) buildingSMART, Certified Software. http://www.buildingsmart.org/compliance/certifiedsoftware/ (retrieved: 18.11.21)
NN (2021au) https://de.statista.com/statistik/daten/studie/157902/umfrage/marktanteil-der-genutzten-betriebssysteme-weltweit-seit-2009/ (retrieved: 28.08.2021)
NN (2021av) https://blenderbim.org/download.html (retrieved: 28.08.2021)
NN (2021aw) https://blenderbim.org/blenderbim-vs-revit.html (retrieved: 28.08.2021)
NN (2021ay) https://www.irbnet.de/daten/rswb/17089005133.pdf (retrieved: 28.08.2021)

5 Data Management and Data Exchange for BIM and FM

Maik Schlundt, Thomas Bender, Nancy Bock, Michael Härtig, Markus Krämer, Michael May, Matthias Mosig and Marko Opić

5.1 Data Management

A key task of CAFM software is to take over, manage and evaluate FM-relevant data (e.g. building data or asset data), which are necessary to support and control facility processes throughout the entire lifecycle of buildings.

The foundation for a valid data basis is already laid in the planning and construction phase (new construction) of a property. The inventory data relevant to FM are ideally generated in accordance with defined rules and standards in coordination with FM in the construction project and can be finally handed over to operations as as-built data from the BIM model at the end of the construction phase.

M. Schlundt (✉)
DKB Service GmbH, Berlin, Germany
e-mail: maik.schlundt@dkb-service.de

T. Bender
pit – cup GmbH, Heidelberg, Germany
e-mail: thomas.bender@pit.de

N. Bock
BuildingMinds GmbH, Berlin, Germany
e-mail: nancy.bock@buildingminds.com

M. Härtig
N+P Informationssysteme GmbH, Meerane, Germany
e-mail: michael.haertig@nupis.de

M. Krämer
Hochschule für Technik und Wirtschaft Berlin, Berlin, Germany
e-mail: markus.kraemer@htw-berlin.de

M. May et al. (eds.), *BIM in Real Estate Operations*,
https://doi.org/10.1007/978-3-658-40830-5_5

For existing buildings that are already in operation but for which no valid digital data basis exists yet, a separate process for inventory data capture has to be implemented. Meanwhile, there are efficient methods and technologies such as photogrammetry or laser scanning (cf. Sect. 5.2) to quickly set up a valid data basis up to the subsequent generation of a digital building model (BIM model).

Inventory data describes a building including its facilities and technical assets. This includes:

- Alphanumeric data, e.g. tables, directories, lists, calculation results or textual descriptions,
- Graphical and geometric data, e.g. parts of BIM models, CAD plans such as floor plans, sections or views, schemas, sketches, building scans, photos and videos.

Inventory data forms the basis for digital processes handling in a CAFM system. In particular, in the case of new construction projects, it is recommended to form a solid data basis as early as possible in the lifecycle. Measures can be:

- Creating structured specifications, including identification systems and required data volume,
- Fixing these data specifications in tender documents and contracts with planners, architects and contractors,
- Continuous checking of compliance with these data specifications and quality assurance up to handover to the operation phase.

Notes on the capture of inventory data are contained in the GEFMA guidelines 420 "Introduction of a CAFM system" (NN 2017a) and 430 "Data basis and data management in CAFM systems" (NN 2019m) as well as the CAFM textbook (May 2018a).

It is recommended to develop an internal data or documentation guideline binding on the company or organization, which should then also be integrated into the EIR for the processing of BIM projects (cf. Sect. 3.4) This should describe both the data structure and the level of detail as well as the responsibilities for data maintenance. The GEFMA guideline 198 "FM documentation" (NN 2013b) can be used to support this.

M. May
Deutscher Verband für Facility Management (GEFMA), Bonn, Germany
e-mail: michael.may@gefma.de

M. Mosig
TÜV SÜD Advimo GmbH, München, Germany
e-mail: matthias.mosig@tuvsud.com

M. Opić
Alpha IC GmbH, Nürnberg, Germany
e-mail: m.opic@alpha-ic.com

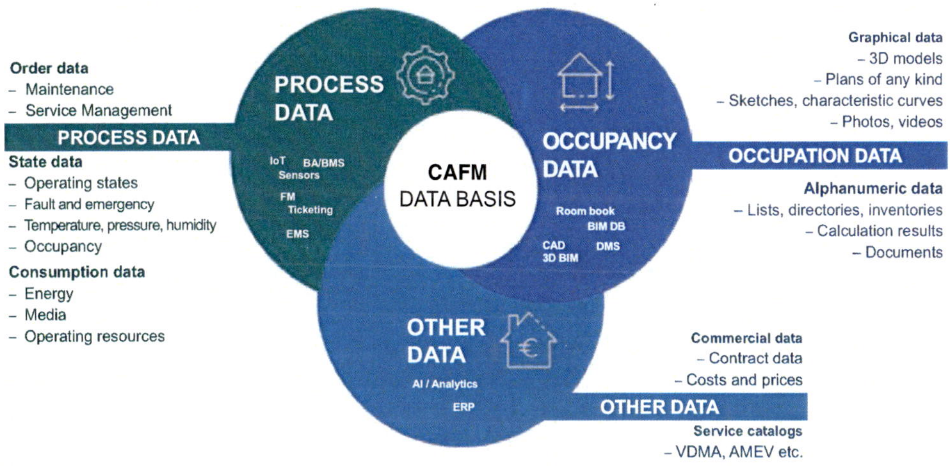

Fig. 5.1 Data categories and their dependencies

Figure 5.1 (NN 2021a) shows the individual data categories and their dependencies.

From the perspective of FM, the data must be exchanged and displayed between the CAFM and BIM software (authoring tool or CDE platform). Both import and export files in different formats (IFC, COBie, CAFM-Connect or other model-based formats, see Sect. 5.3) must be readable with suitable viewer software so that their content can be displayed and checked in a structured way.

The following functions must be provided for process support by CAFM software:

- Transfer of a spatial structure including data from the BIM model into the database of the CAFM software,
- Display of the structure and data in the CAFM software,
- Transfer of space data from the BIM model into the CAFM software,
- Transfer of the technical asset data from the BIM model into the CAFM software,
- Visualization of graphical and geometric data from the BIM model within the CAFM software,
- Transfer of equipment/inventory with their essential characteristics from the BIM model into the CAFM software,
- Transfer of changes in the CAFM software (data, possibly also geometric data) back into the BIM model or the BIM authoring tool,
- Import of IFC files into the CAFM system and export of IFC files after editing,
- Import/export of COBie or COBie-Lite files,
- Check and display of the files to be exported,
- Transfer and further processing of events from the CDE software (e.g. defects).

The following data and catalogs must be processed:

- Buildings, floors, rooms with room numbers and room types,
- Room area with type of use according to measurement standards such as IPMS or DIN277,
- Floor plans,
- Technical assets, inventory, furniture each with manufacturer, model/type number, characteristic data, etc.

The following reports should support the decisions in facility management:

- (Error) protocols of data transfer,
- Technical asset list,
- Room book,
- Visual object representation.

Interfaces to BIM software (authoring software, CDE software) are useful for an extensive system integration (see Sects. 4.2 and 4.3).

A special challenge in the exchange of data between CAFM software and BIM software (authoring software, CDE software) arises from the fact that, based on the data relevant to FM, different data models are used in both software environments. Therefore, different initiatives have emerged on the market to ensure a standardized data exchange.

Most of these initiatives are based on the standard exchange format IFC for BIM authoring software, with additional FM-relevant parameters that can be added. An overview of the initiatives that currently provide data models for the exchange of data between a BIM model and a CAFM system can be found in Appendix 2. However, the FM-relevant parameters are usually not standardized across the board and therefore not always coordinated with each other. This often leads to gaps in the transfer of FM-relevant parameters, e.g. for cleaning management or for inspection requirements for technical assets.

A leading standard for the extension of FM-relevant parameters in a BIM model has not been established yet. Rather, initiatives are currently emerging that can map the existing various standards to each other (data mapping), in order to enable platform-based data exchange even across platforms (see Sect. 10.2.1).

The recommendation is therefore to define the data attributes relevant for operation from the target system (often the CAFM system), to reconcile them with an IFC-based standard and then to close the gaps identified there by user-defined settings. The resulting data schema, including unique reference/identification keys for spaces, inventory features or technical assets and their hierarchical order, must then be attached as an annex to the EIR and agreed with the contractors of the BIM project as the contractual basis.

CAFM software providers work with vendor-specific plugins based on proprietary data formats (usually Autodesk Revit, see Sect. 4.3.2.2) or even the open IFC format. Both unidirectional interfaces with a simple IFC import, but also a bidirectional exchange (IFC import and export) is possible.

However, data management in operation poses another challenge for many companies. Changes to inventory data as part of conversions, changes to room layout, renovations, larger repairs, etc. are often not updated centrally, which means that inventory data is no longer up-to-date and complete (see Sect. 4.3.2.3). As a result, the process data in the CAFM system is no longer based on consistent inventory data and distorts the results. For example, inventory data must be structured, verified and re-entered every 3–5 years when FM services are put out to tender. To avoid this, a continuous data management process must be implemented in the operational phase including the resources required (internal/external employees and tools). This process should at least cover the following activities:

- Identification of changes to inventory data as part of service provision (e.g. conversion) or by special inspection,
- Communication of changes to inventory data as part of service documentation or a separate workflow,
- Complete and up-to-date digital documentation of changes in one central location (e.g. CDE, DMS, digital project space),
- Quality assurance of the update in accordance with the documentation policy (e.g. EIR for inventory changes) and
- Update of the links between these inventory data and the CAFM system.

5.2 Modern Data Capture for BIM and FM

A major challenge for the use of the BIM method in the practice of FM arises from the fact that, as a rule, for existing buildings, which make up the majority of the real estate market, no digital building models (BIM models) are available. BIM models can only be used meaningfully in FM if they also reflect the current state of the facilities, that is, if they are based on *as-built* information. The effort required to create such digital as-built models is currently considered too complex by many FM organizations.

The following sections therefore focus on digital data capture methods that can be used as a partial replacement or supplement to a manual, subsequent modeling of an existing building. Further methods of data capture for existing buildings, in particular for capturing HVAC components, are explained in detail by Turianskyj et al. (2018).

5.2.1 BIM Modeling for Existing Buildings

The creation of BIM models of existing buildings for FM goes far beyond the usual process of building survey. While 2D CAD drawings such as floor plans, views or sections were predominantly in the foreground, the creation of BIM models requires not only the creation of mostly volume-oriented 3D geometries, but above all a component-related 3D object creation, including the technical object attributes (object semantics) and object classification. If, for example, a door is created using a BIM authoring system, this includes the (simplified) 3D geometry of the door, the assignment of a component class (in the case of the IFC format, for example, IfcDoor) and the door properties considered important for operation (for example, the fire protection class).

However, classical 2D plans can also form the starting point for model creation. However, the prerequisite is that all plans represent the current state of the building and thus also subsequent structural adaptations have been carried out. If this is not the case, the usual capture methods, such as an electronic distance meter or total station can be used (Turianskyj et al. 2018). However, the actual modeling activity is carried out without automated object recognition.

In order to reduce the modeling effort, the geometric detail level (Level of Geometry—LOG) or the alphanumeric detail level (Level of Information—LOI) can be adapted to the tasks in FM or operation (cf. Sect. 3.4). The advantage of such manual modeling based on existing and updated 2D plans is the comprehensive information of classical plans, which goes far beyond the pure geometry information and, for example, contains information on materials used or room stamps.

5.2.2 Digital Capture Methods for the Building Documentation of Existing Buildings

However, if no 2D plans are available or if they are too outdated, surface- or volume-oriented digital capture methods such as terrestrial 3D laser scanning or photogrammetric methods can be used. The aim is to reduce the amount of data capture and, if possible, the amount of modeling effort, but this is not always possible. For this reason, it is crucial to use the capture results as efficiently as possible. Section 5.2.4 explains different scenarios that pursue this goal.

5.2.2.1 Terrestrial 3D Laser Scanners

Terrestrial 3D laser scanning (light imaging, detection and ranging—LIDAR) not only offers a fast and accurate point measurement, but also allows surveyors to collect data up to one kilometer away (depending on the type of equipment). The end result of a 3D laser scan is a precise and dense set of measured points, which together are referred to as a 3D point cloud. In the point cloud, each point has x-, y-, z-coordinates (or can be converted into the Cartesian coordinate system), where depending on the method the time of flight of the laser beam emitted by the laser is measured. Such a point cloud usually comprises millions of individual points, which can be recorded in a few minutes with an average point accuracy in the range of a few millimeters by current devices. It should be noted that a 3D laser scanner can only capture the visible area of a building, that is, the areas that can actually be reached by the laser beam. Objects behind suspended ceilings, installations inside walls or other objects behind barriers cannot be captured by this method.

5.2.2.2 Photogrammetric Methods with Surveying Drones

In addition to 3D laser scanners, building data is also captured by photogrammetry. These methods are applied, for example, when using surveying drones. Although surveying drones can also be equipped with LIDAR sensors, video- or photo-based methods are usually preferred. In photogrammetric methods, a large number of individual images are captured, which are then converted into a 3D point cloud by photogrammetry software at a later stage, with the exact position being determined for each individual image. For the capture of a building, a surveying drone takes about 1000 individual images in a flight mission of about 10 minutes, which are then transformed into a 3D point cloud. However, such 3D point clouds have a significantly lower accuracy than the result of a 3D laser scanner (see Krämer et al. 2017). Since a surveying drone is usually programmed to fly a flight path at a constant height, for example, in which a building is surveyed following a meandering trajectory at a height of 30 m, the accuracy decreases significantly due to the capture angle of the camera, especially in the lower part of the building.

5.2.3 Workflow for BIM Modeling with Digital Capture Methods

This section explains the major workflow of creating a BIM model for existing buildings —related to the two digital capturing methods explained above (Krämer and Besenyöi 2018). The workflow in Fig. 5.2 includes the individual activities as well as their time expenditure, but also provides information on the importance of the activity for the overall result. Furthermore, it includes the mapping of three scenarios for using a point cloud: *Scan2CAFM*, *Scan2Dataset* and *Scan2BIM*, which are explained in Sect. 5.2.4. The basic idea of the three scenarios is not to transform the complete 3D point clouds into a parametric BIM model in any case. A further utilization of the point clouds is quite sufficient for certain tasks in real estate management. In this way, a so-called hybrid BIM model is created, which comprises parametric building elements (such as rooms, walls, doors, windows) and associated point cloud segments as well as alphanumeric data in a CAFM system. The aim of these hybrid models is to reduce the modeling effort while maintaining a sufficient level of information (BIM Lite).

In the first phase (data acquisition), the focus is on the correct positioning of the equipment (scan stations). The definition of the positions of the scan stations ensures that

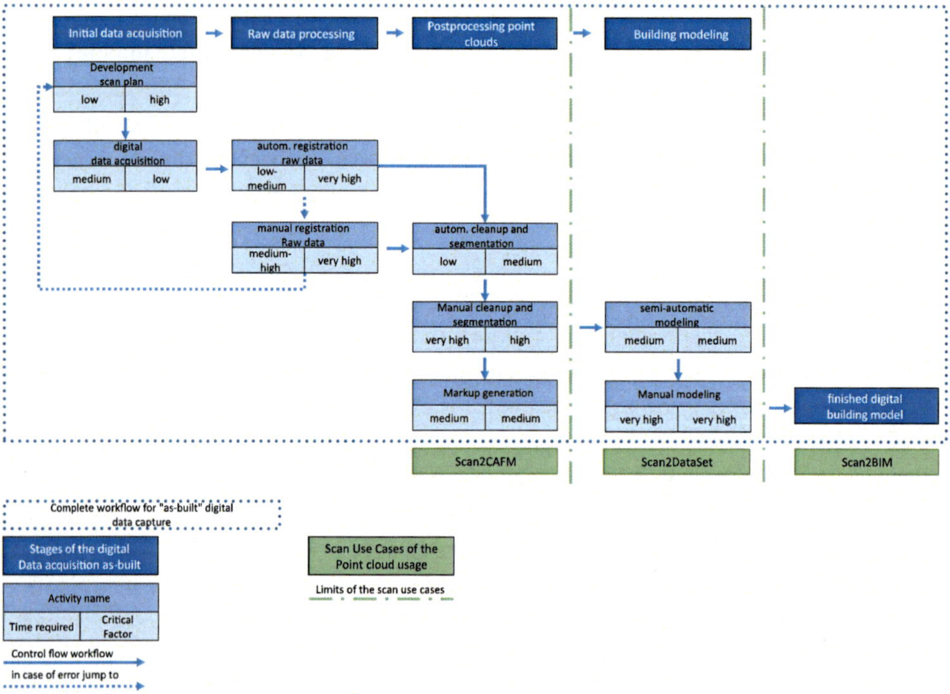

Fig. 5.2 Workflow for the creation of BIM models for existing buildings

the data (raw data) collected have sufficient overlap with each other. This ensures further processing of the raw data. For example, 25% of the points collected must also be included in the scans of the adjacent scan stations. This might be achieved by ensuring the scanning device is positioned in each door opening, allowing to connect different areas of the building (and rooms) in the individual scans. Accordingly, in the case using a surveying drone, the aerial images automatically taken must also contain overlapping point information of adjacent images.

This activity does not take long (time required: low), while the importance of this activity for the final result is high (critical factor: high). Although the actual data acquisition throughout the workflow has a moderate time requirement, the *critical factor* is only low compared to the proper positioning of the equipment.

In the second phase (raw data processing), raw data is combined according to the overlapping point regions, whereby a common, registered point cloud is created from the individual point clouds of each scan station. This registration process can be carried out automatically, manually or in combination of both methods.

In the case of the third phase (data processing), unnecessary points (noise) are removed from the point clouds that are divided into segments (cf. Fig. 5.3). This phase can also be carried out automatically or manually. Although most of the unwanted points

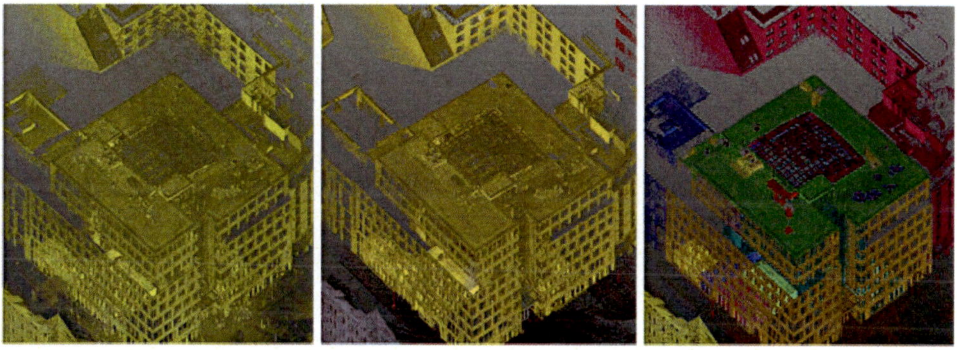

Fig. 5.3 Post-processing of point cloud data

can be removed during automatic cleansing, unwanted points such as silhouettes of people remain sometimes. Segmentation enables working with smaller data sets of the point cloud during the phase of model creation, which significantly increases processing speed. In addition, individual segments can be named and provided with additional information (markups) and can then directly be assigned in a CAFM system, for example, to rooms. It is also common to map technical objects (e.g. assets) as point cloud segments and provide them with descriptive markups (points of interest—POIs) (Scharf 2016).

In the final phase (BIM model generation), the digital building model is created based on the captured point cloud data. This modeling step can be carried out semi-automatically by suitable software or manually using typical BIM authoring systems. If the first option is used, the effort can be reduced significantly. Based on a generated surface model the Edgewise software, can detect building components automatically and transform them into parametric Revit objects. This works quite reliably for objects with simple geometry, such as walls, openings, beams, columns, and pipe elements (see Fig. 5.4). Depending on the method used, however, considerable manual follow-up work is sometimes necessary.

The generation of point clouds using 3D laser scanners can be carried out quite fast on site with today's systems. However, for a building survey, as is usually carried out by FM service providers for new contracts, the effort is still high and the result is often incomplete. Here, photogrammetric recording maybe an alternative (if the accuracy requirements are lower). The use of surveying drones is also an option, especially for difficult-to-access roof areas or the surveying of outdoor areas.

To automate the workflow presented, combined capturing devices such as the M6-Trolley or the portable mapping system NavVIS VLX of the company NavVis are used very successfully, which combine photogrammetric methods, LIDAR recording, 360° panoramic images, and other sensors (Rust and Och 2021). The surveying systems also use an automated data preparation and further processing, which now also supports

Fig. 5.4 Automatic building component detection (left) and manual building modeling (right)

the final step, i.e. the transformation of the point cloud models into parametric BIM objects. In addition, AI-based methods for the automatic detection and transformation of point cloud data are the subject of current research activities (see Sect. 10.2.2).

If the effort for digital capture by building scans has been made, the best possible use of the scan data is one of the scenarios described in Sect. 5.2.4.

5.2.4 Scenarios for the Use of 3D Point Clouds

5.2.4.1 Scan2BIM

The application case *Scan2BIM* includes the complete transformation of the 3D point clouds into parametric component objects of a BIM model of an existing building. This completes the entire workflow from Fig. 5.2. In addition to the time-consuming direct (manual) modeling of BIM objects based on point coordinates, the modeling effort can be reduced by contour-related and semi-automatic methods for object recognition. Great progress is currently being made in the field of automatic object recognition in point clouds by the use of *Artificial Intelligence* (AI) and *Machine Learning* (ML) (see Sects. 2.9 and 10.2.2). However, component data must be supplemented manually in any case. An advantage of this scenario is that a BIM model is created that can be completely transferred to or linked with a CAFM system, as explained in Sect. 4.3. The resulting BIM model of the existing building does not differ basically from a BIM model that is handed over from a new construction project to FM. The disadvantage is certainly the considerable modeling effort. Otherwise serious restrictions in the level of detail must be accepted.

5.2.4.2 Scan2CAFM

The aim of the second application case *Scan2CAFM* is determining the asset information required for FM from the point cloud. The basic idea is to transfer asset data directly into a CAFM system by identifying objects in the point cloud (e.g. building assets) or by taking measurements. In this way, the detour via the modeling of building objects can be omitted. For example, room dimensions and areas can be taken relatively easily. However, the identification of objects or the recognition of scanned machine plates is not possible or only possible to a limited extent due to the low resolution of scanners. However, great progress is to be expected in this area in the future. The scenario *Scan2CAFM* should be used in combination with the scenario *Scan2Dataset* explained below. Hence, the point cloud information can be utilized in operation beyond the extracted asset data and measurements.

5.2.4.3 Scan2Dataset

The scenario *Scan2Dataset* can be combined with the two scenarios explained before. The basic idea of this scenario is to reduce the effort during initial BIM modeling by distributing the modeling or detailing of objects over the operating phase on-demand. For this purpose, the point cloud segments generated in step 5 of the workflow in Fig. 5.2 (e.g. a room or asset) are provided in point cloud files via a database. These point cloud segments can then be assigned to a floor or a room in a CAFM system. In this way, the room is initially not represented as an object in the BIM model (cf. *Scan2BIM*), while important information, e.g. for checking the installation space or accessibility of an asset, can be directly determined from the point cloud. A transformation into BIM objects is not necessary, but can be carried out on demand if required. The combination of the scenarios then corresponds to a hybrid BIM model.

5.2.5 Other Methods

Another application for increasing the efficiency of controlling of new construction or renovation projects is an AI-based superimposition of building scans (or photos) with the BIM model to detect deviations. This makes it possible to obtain an up-to-date and complete as-built documentation from the construction project for the operational phase. For this purpose, the state of the project is captured periodically on the construction site by mobile scanners or using photogrammetry. The results of the point clouds (or photos) are then superimposed by the BIM model from the execution planning and analyzed using artificial intelligence. Deviations, for example with regard to the existence or position of technical components, are thus identified and marked as defects. These defects can then be classified. If serious deficiencies are detected, they are fed into the defect tracking process. In this way it is ensured that the building, including the assets, corresponds to the BIM model. Conversely, technical changes required on site can be identified and considered in the as-built model.

5.3 Methods and Formats for BIM Data Exchange

5.3.1 Industry Foundation Classes (IFC)

IFC is an open, vendor- and platform-independent object-oriented data model standard for exchanging data in the construction industry and beyond. It usually forms the basis for Open BIM projects (see Sect. 3.4.4). IFC was developed by buildingSMART International and has been registered as an international standard ISO 16739:2018 (NN 2018a) since release 4. Several hundred classes describing real objects are already included in the current version. Within the framework of IFC, all components that exist on or in a building are defined as objects. In software programs supporting this interface, these objects can be loaded, further processed and edited. For example, a window of a building can be exchanged via this interface with all its properties and related information.

The platform independence and the resulting data exchange between different software systems support integrated work of the partners involved in the processes. Furthermore, time and costs are saved while increasing quality. With the product data model of IFC, an attempt is made to enable process optimization in construction by interlinking all project participants, such as architects, engineers, contractors, building services engineers and facility managers, by the same data basis.

Specifications, documentation and implementation guidelines for IFC are freely available and provide an object-oriented data schema (NN 2021ap). This format is used to transfer/exchange, export and import data between different software applications. The IFC's define an integrated, object-oriented and semantic model of all components, attributes, properties and relationships within the product "building". This means that all components and properties are described by IFC as objects and interpreted as such within IFC-compatible software. Components and properties not only refer to building elements, rooms and geometric shapes, but also to logical, topological and temporal relationships.

With the IFC standard, objects remain unchanged with their properties throughout their life cycle and can be extended by additional information, e.g. from the administrative sector or FM. This makes it possible to store not only geometry, but also costs, heating loads or reference documents such as drawings or images in IFC-compatible software and to process them further with different software systems.

IFC-compatible software supports one or more data exchange processes. Examples are:

- Exchange between two disciplines in the same design phase
  (e.g. for coordination of the breakthrough planning between the CAD systems for architecture and MEP),
- Exchange within a discipline in two design phases

(e.g. for transfer of the preliminary design from one architecture CAD system to another, used by another office for execution planning),
- Exchange within a discipline in one design phase
(e.g. for transfer of models of a structural modeling system to a structural analysis program),
- Exchange within two lifecycle phases and two disciplines
(e.g. for transfer of the construction documentation from the architectural planning to the CAFM system of the operator).

For these data exchange processes, IFC-compatible applications implement the required subset of the IFC data model. These subsets are referred to as "Views" (model view definition—MVD). An MVD is a view of a data model—in this context a view of the product data model of IFC (NN 2021aq). MVD describes the classes of the product data model, which must be supported by the software in order to ensure the corresponding data exchange processes. IFC-compatible software systems are usually certified. The implementation and certification always relate to an MVD.

A software can also implement several MVD′s, but each MVD must be certified separately. In order to ensure successful data exchange, both the sending and receiving system must support the same IFC version and MVD.

However, the transition from one IFC release to the next is often still accompanied by problems, because older versions are kept longer than is reasonable. It is also important to keep in mind that the versions are usually not backward compatible, for example, the current IFC4 format is not backward compatible with the much more widely used IFC2x3 format.

The IFC format is continuously developed by buildingSMART. One main problem is that IFC was created for data exchange based on files, and is therefore not optimized for new applications such as database-oriented CDE's for representing digital twins or connections to IoT. New developments must therefore consider these upcoming requirements.

5.3.2 BIM Collaboration Format (BCF)

BCF is an open file format that was developed by buildingSMART. It is used for platform-independent communication of IFC models between all parties involved in the various phases such as planners and specialists. Using texts or images, a model can be commented on or, with reference to individual model elements, a discussion can be supported. These data are stored as BCF files or exchanged via a web API. The file format is independent of the IFC file. However, the contents to be transported in the BCF file are assigned (referenced) to the elements within the IFC file. This is done using the globally unique identifier GUID specified in the IFC standard. The file is created in XML format (bcfXML). The technical documentation is freely available from buildingSMART.

BCF allows a model-based communication across different BIM software applications such as authoring tools and viewers, because comments and annotations can be edited in the viewer as well as in other software for enriching BIM data. Typical applications are the support of logging, documentation and controlling collaboration and change management (e.g. issue tracking).

5.3.3 Construction Operations Building Information Exchange (COBie)

COBie is a data exchange standard for non-graphical BIM data (NN 2021ar). This exchange format was developed by the American military.

COBie can be used to provide data that plays a role during operation, maintenance and repair, as well as asset management. Data capture can extend across the entire lifecycle of a property, that means from planning and construction to operations.

COBie is originally a tabular format in which building information is provided in the form of alphanumeric attributes. Opening and editing a COBie file is usually done by a spreadsheet program. COBie offers 16 pre-formatted data sheets that are linked together. Information such as building structure, MEP and documents can be described in a structured way. Therefore, a BIM tool is not necessarily required for data maintenance.

If a CAFM software has a COBie interface, the basis for successful import of COBie-based BIM models is given. However, it should be noted that COBie does not describe contents such as those of an asset identification system or general performance data of an object. These should be defined in advance.

Examples of data to be captured and exchanged are:

- Equipment and inventory lists,
- Spare parts lists,
- Product data sheets,
- Information on warranties and
- Maintenance schedules.

COBie uses various formats, such as spreadsheets, IFC or ifcXML, to exchange data. COBieLite is a simplified XML structure introduced by buildingSMART. All of those can be used for standardized data exchange.

5.3.4 Green Building eXtensible Markup Language (gbXML)

gbXML enables the interoperable exchange of data between computer aided architectural design systems (CAAD) and technical calculation programs or analysis tools (NN 2021as). It is an open format and consists of the simple building geometry in XML format, through which room data including the building envelope can be exported and thus complete room books transferred into calculation programs. This eliminates the need for manual re-entry and new input into calculation programs. The gbXML format is used, for example, to export the data required for energy calculation from a BIM model.

5.3.5 CAFM-Connect

CAFM-Connect is a German initiative to ensure the interoperability of software used in CAFM. It forms a uniform interface (import/export) for the exchange of alphanumeric data between FM data capture and CAFM systems.

CAFM-Connect enables easy import of basic FM data into CAFM systems that support this exchange format. At the same time, it is possible to export FM data into systems that require these data for successful operation and support IFC4 or the predecessor IFC2x3. CAFM-Connect 1.0 included space data that was hierarchically divided into property, building, floor, and room, while CAFM-Connect 2.0 also included asset and equipment data. The current version 3.0 also classifies document types and supports, among other things, the storage and transport of documents (files) based on a classification according to GEFMA guideline 198 and the IFCZIP file format (NN 2021f).

5.3.6 Proprietary Exchange Formats

In addition to open formats that are usually internationally standardized, of course, various BIM software providers use proprietary formats. This can be particularly useful if BIM projects are carried out within the product family of a BIM software (Closed BIM). On the other hand, various CAFM software vendors offer interfaces to such proprietary formats. The most prominent example of such a format is the Revit format rvt, which contains the complete building information model with all geometric and textual subject data as well as meta-information and can also be stored in an SQL database. Therefore, it is more of a proprietary database than a classic file format. At the "IT in Real Estate and Facility Management" conference in 2017, the integration of IT systems from different manufacturers into a single workflow was impressively presented live on this basis as part of the "CAFM Future Lab" (May 2018a).

5.4 BIM-FM Data Manager

In the planning and construction phase, there should be clear roles such as the BIM information manager (cf. Sect. 7.2), which are contractually obliged by the EIR and the BEP to also transfer the FM-relevant data in the form of an as-built model to the operation and the CAFM software.

On the operational side, a BIM-FM data manager should be introduced at this interface for data transfer and further updating in case of changes during operation. This role can be filled internally or externally and is responsible for the fact that the inventory data of the buildings are up-to-date and complete and centrally accessible to all stakeholders. If outsourced externally, it is important that the scope of services is described clearly, as there is no standard service description for this service and there are currently only a few providers specializing in this role.

These services are often provided by external regional architects or surveying offices or the external FM service provider is commissioned. When awarding the contract, attention should also be paid to the technical competence of the provider for the maintenance of the MEP inventory data. It is also critical that the client always has full legal access to their data.

The BIM-FM data manager establishes, controls and executes a data management process that must include the following sub-services:

- Formulation, updating and integration of data responsibility manuals into construction projects,
- Establishment of quality assurance processes for compliance with the requirements in the projects,
- Takeover of existing data in the form of BIM models, CAD data, alphanumeric master data and digital documents,
- Carrying out quality assurance of these data,
- Communication of defects and initiation of defect removal,
- Import or linking of these data into the CAFM software,
- Establishment of a change process in the operational phase,
- Recording of changes to existing data in the operational phase,
- Classification and qualification of these changes,
- Organization or carrying out data transfer or data capture after changes,
- Updating of graphical and alphanumeric data in the CAFM software and
- Transfer of existing data to the project team in the event of conversion, refurbishment, renovation.

Depending on their own level of performance, sub-services can be carried out internally or externally. It is important to note that there are clear, standardized requirements for all parties, projects, measures and involved software systems that ensure compatibility when transferring and exchanging data.

As a tool for the data manager to check compliance with the requirements in BIM models, the use of model checkers is recommended (see Sect. 4.2), which can check the completeness of the parameters, compliance with nomenclatures and the existence of required model elements. The number and size of the annual projects and change measures and the complexity of the operational processes dependent on these data determine the internal or external staff capacities for this role. For example, airports sometimes have whole departments for updating the inventory data, while in office buildings with around 2500 employees this role can also be carried out in synergy with the role of the space manager (depending on the office space concepts).

The following qualifications and skills should ideally be available when assuming this role:

- Basic understanding of the planning and construction process,
- Operation of CAD and modeling software,
- Operation of CAFM software,
- Operation of CDE software or the digital project space (if available),
- Knowledge of the FM-relevant data structures (architecture, technology, equipment, etc.),
- Control of external service providers and
- Possibly configuration of the rules and operation of model checkers.

Thus, the BIM-FM data manager is the interface for transferring BIM data to operation.

5.5 Summary

The focus of this chapter is on the data necessary for applying the BIM methodology in real estate and FM. The basics of data management are introduced. For example, it is discussed which type of data is needed. The data basis for the BIM method ideally consists of as-built models. If this data is unavailable, which is often the case with existing buildings, it is possible to re-enter the data. 3D laser scanning is suitable for this, with the respective methods of post-processing such as SCAN2BIM, SCAN2CAFM or SCAN2Dataset.

For further processing and exchange of BIM data, various formats and methods have been developed. These include standardized formats such as IFC, COBie, gbXML and CAFM-Connect, which support the Open BIM approach. In addition, there are proprietary formats that are often used as a standard format for a vendor's software. The different formats and their uses are described.

The process is concluded describing the role of a BIM data manager, who is responsible for coordinating the data management process and the data from an FM perspective. In addition to creating imports and exports, his tasks also include checking the data quality.

References

Krämer M, Besenyöi Z (2018) Towards Digitalization of Building Operations with BIM. IOP Conference Series: Materials Science and Engineering, IOP Publishing Ltd, Moskau, pp 1–11

Krämer M, Besenyöi Z, Lindner, F (2017) 3D Laser Scanning – Approaches and Business Models for Implementing BIM in Facility Management. Proc. INservFM, Verlag Wissenschaftliche Scripten, Auerbach/Vogtland, pp 679–691

May M (Ed.) (2018a) CAFM-Handbuch – Digitalisierung im Facility Management erfolgreich einsetzen. 4. edn., Springer Vieweg, Wiesbaden, 2018, 713 p

NN (2013b) GEFMA 198: FM-Dokumentation, November 2013

NN (2017a) GEFMA Richtlinie 420: Einführung von CAFM-Systemen, Juli 2017, 7 p

NN (2018a) DIN ISO ISO 16739-1: Industry Foundation Classes (IFC) for data sharing in the construction and facility management industries Part 1: Data schema. International Organization for Standardization, 2018-11

NN (2019m) GEFMA 430: Datenbasis und Datenmanagement in CAFM-Systemen, März 2019, 10 p

NN (2021a) GEFMA Richtlinie 400: Computer Aided Facility Management CAFM – Begriffsbestimmungen, Leistungsmerkmale, März 2021, 19 p

NN (2021f) https://www.cafm-connect.org/ (retrieved: 27.05.2021)

NN (2021ap) https://www.buildingsmart.org/standards/bsi-standards/ (retrieved: 17.08.2021)

NN (2021aq) Model View Definition. https://technical.buildingsmart.org/standards/ifc/mvd/ (retrieved: 17.08.2021)

NN (2021ar) https://cobie.buildingsmart.org/history/ (retrieved: 17.08.2021)

NN (2021as) https://www.gbxml.org/ (retrieved: 17.08.2021)

Rust C, Och S (2021) Den digitalen Zwilling erzeugen. Bauen im Bestand 44(2021)5, 64

Scharf H-J (2016) Panoramabilder, Punktwolken und Points of Interest. Der Facility Manager 22(September 2016)9, 36–37

Turianskyj N, Bender T, Kalweit T, Koch S, May M, Opić M (2018): Datenerfassung und Datenmanagement im FM. In: May M (Ed.) CAFM-Handbuch – Digitalisierung im Facility Management erfolgreich einsetzen. Springer Vieweg, Wiesbaden, pp 229–258

6 Economic Efficiency of BIM in FM

Markus Krämer, Thomas Bender, Matthias Mosig and
Marco Opić

6.1 Drivers for Value Creation through BIM

In addition to the GEFMA guidelines 400ff and with reference to RL 460 (NN 2016a), some definitions in the field of economic efficiency are presented below (NN 2016a).

Economic efficiency

Economic efficiency is described in the literature as the optimal ratio between input and output. An IT system is considered economically efficient if the costs of implementation and operation are below the measurable benefits expected within the period under consideration.

M. Krämer (✉)
Hochschule für Technik und Wirtschaft Berlin, Berlin, Germany
e-mail: markus.kraemer@htw-berlin.de

T. Bender
pit – cup GmbH, Heidelberg, Germany
e-mail: thomas.bender@pit.de

M. Mosig
TÜV SÜD Advimo GmbH, München, Germany
e-mail: matthias.mosig@tuvsud.com

M. Opić
Alpha IC GmbH, Nürnberg, Germany
e-mail: m.opic@alpha-ic.com

M. May et al. (eds.), *BIM in Property Management*,
https://doi.org/10.1007/978-3-658-40830-5_6

ROI

The profitability or the return on investment (ROI) compares the profit from an investment with the capital investment. However, as a static indicator in investment accounting, ROI does not take into account the different courses of deposits and withdrawals within different periods (e.g. years) of a period under consideration. In common usage, ROI is often understood to mean the profitability of an investment in general, without specifying in detail under which conditions it is to be calculated.

In order to determine the economic efficiency, three steps have to be carried out essentially:

- Determination of the expected costs,
- Determination of the achievable benefits and
- Calculation of the economic efficiency.

The individual value drivers can influence both the costs and the benefits. If considering the drivers for the value creation through BIM throughout the lifecycle, a hierarchical tree of drivers is resulting. It starts with the production and can be further aggregated until disposal.

The drivers can have an influence on the following aspects:

- Reduction of process cycle times, e.g. through savings in working time as a result of increased efficiency among all parties involved,
- Reduction of costs through EBIT-effective savings in external material costs through, for example, improved planning quality and the avoidance of error costs, a need-based design of architecture or MEP,
- Rising sales and profits through increasing value of the property and a higher sales price. This can be achieved, for example, through complete and up-to-date BIM-based inventory documentation, through avoiding the risk of a delayed start of core business activities or through preventing the loss of rent income due to accomplished target dates.

The drivers can have a direct or indirect influence on value creation of the respective phase regarding expenditure and benefit considerations. For example, manufacturers can use BIM object and product platforms to expand direct sales channels in the future or contribute to information management of sustainability objectives (tracking of production chain). A detailed description of the application of a driver model for CAFM systems can be found in GEFMA 460, which can be transferred to BIM applications to a large extent (NN 2016a, see also May 2018a).

6.2 Economic Efficiency of BIM in the Construction Phase

The profitability of BIM is determined in the construction phase essentially by the comparison of investments for technical infrastructure necessary for implementing the BIM method (e.g. server, software, platforms) as well as for training, qualification and integration of the roles involved in the BIM project (e.g. the role of the BIM manager) and the respective benefits. The benefits arise mainly from savings due to the avoidance of planning errors, subsequent work, shortened planning and construction times and an optimized operating phase which is explained in terms of value creation in Sect. 6.2.1 to 6.2.3.

6.2.1 Value Creation Related to Process

BIM avoids unrecognized planning errors in the planning phase and construction errors caused by later misinterpretations of plans during construction phase. It is important that the BIM method is applied in the sense of a *Digital Prototyping* and not only understood as a modern form of digital documentation. Although the simple documentation of construction defects already works better with BIM than with 2D plans, the immense additional potentials are only unlocked by a prototyping approach enabled by the BIM method. The resulting avoidance of planning and construction errors is also reflected in the operational phase, in which errors lead to increased consumption or unused resources. If suitable countermeasures are initiated in time as a result of the application of the BIM method, the benefits go far beyond a simple documentation.

Many clients today believe that a construction-accompanying planning process with all its risks such as significant time and cost increases cannot be avoided because construction projects are under high time pressure. However, the practice of using the BIM method shows that, when implemented consistently in the project in the sense of a *Digital Prototyping* and a *Digital Lifecycle Management,* a considerable time saving can already be achieved in the design and planning phase by avoiding time-consuming (re-) planning after completion of Phase 5 according to HOAI. The term digital prototyping refers to the simulation of aspects such as energy, area, (maintenance) processes (e.g. accessibility for maintenance purposes), which are made possible by the use of the BIM (cf. Sect. 2.11). BIM is therefore also a risk reduction method, from which a high potential for savings arises.

The key to the value-creating use of BIM is the implementation of BIM already in the pre-project phase or in the design phases. With BIM, we no longer just plan, but the BIM method enables to digitally construct a building prototype, assuring its performance over the lifecycle by simulation. This, for example, leads to the optimization of maintenance intervals or cash flow in a 5D BIM planning.

6.2.2 Quality-Related Value Creation

Even new buildings can usually not be referred to as "new" at the time of completion, since the installed technical components have shorter life cycles compared to the duration of the planning and procurement phase in construction projects. During a 2–3 year planning process, often several generations of technical components are developed in parallel, which have a better performance, sustainability or energy balance. So far, these new components could not be considered in an advanced planning stage, not least because the required simulations had to be manually programmed to prove savings effects and were therefore usually more expensive than the expected added value. As a result, often old technical components are not replaced by more energy-efficient subsequent generations in the planning in order to avoid manual recalculation and adaptation of complete trade or component lists.

By using BIM-based software and its parametrized ("intelligent") digital BIM component objects, technical components can also be simulated, evaluated and easily replaced during the period between planning and execution or even during the execution phase. The use of appropriate BIM tools in the planning and execution phase then results in an improvement of the building performance as a quality-related value creation contribution.

With BIM-based simulation, end-of-life scenarios and material scenarios can also be checked more easily. With specific software programs, for example, materials or different versions of the future building can be compared with each other in a benchmarking. For example, it can be compared how the acquisition, installation, cleaning and disposal costs of different materials such as linoleum, oak parquet or granite floor behave.

6.2.3 Resource-Related Value Creation

6.2.3.1 Space Efficiency

What is not needed in terms of space does not need to be built and maintained. For example, the planning of an international medical center was optimized by means of BIM simulation such that the technical area for facilities could be reduced from 25 to 7% while ensuring comparable maintainability.

This also applies to the undersizing of technical areas, since architects often do not adequately consider the aspect of accessibility of these areas when planning the areas for HVAC installations. By simulating maintenance space with respect to their facilities and components, the building can be plausibilized with regard to its operational space requirements already during the planning phase. This enables optimization of operations later on. For example, a European hospital was planned throughout with a focus on optimal operations, resulting in considerable savings each year through reduced access routes resulting in time savings and thus reduced personnel resources for technical FM.

This was only possible because facilities are quickly and easily accessible during operations.

6.2.3.2 Energy Efficiency

By using intelligent BIM objects and their contribution to improved analytical ability, buildings can be better optimized with respect to their overall energy balance using BIM. In this way, buildings are created that match the energy requirements of the National Action Plan Energy efficiency (NAPE) of the BMWi (NN 2019h), resulting in energy savings at the macroeconomic level. This also applies to lifecycle costs, which can be simulated up to demolition, so that buildings can also be optimally adapted to environmental policy requirements with regard to used materials. In the past, planners often did not recognize the consequences of using a parquet or linoleum floor in terms of cleaning, sustainability, deconstruction and recycling.

In comparison to the previous approach, BIM enables stakeholders to use simulations with a much lower-threshold and thus more often and to avoid the expensive programming of individual simulations, as was necessary in the past. In addition, at least to some extent, the need for specific mathematical and physical knowledge, which was previously required to develop and apply such simulation methods, is eliminated. Extended program logic and new simulation features enable today simulation of even complex scenarios much faster and more effeciently, so that buildings can be optimized energetically but also with respect to many other parameters today.

Model-based calculations (cf. Sect. 2.11) save a considerable amount of communication time between CAD/BIM construction and engineering, which benefits project time and thus costs. In the portfolio, further savings can be generated by using BIM models as the basis for a "realtime building dashboard" for analyzing and optimizing existing buildings.

6.3 Economic Efficiency of BIM in the Operational Phase

Economic efficiency of BIM in the operational phase can be demonstrated by numerous BIM use cases, which however depend on the specific conditions of the drivers for costs and savings. Input variables, such as the size of the property and the property portfolio, the number of employees for operations, the complexity of the operational processes, the chosen operator model and the frequency of relevant events such as moves or service requests, are decisive for the respective benefits. A first overview of the expected benefits of BIM use in FM is given in Sect. 3.3 based on commissioning, executed operational processes and renovation or conversion during operations. In the following, some BIM use cases will be elaborated with regard to the benefits of reducing process times, saving material costs and increasing the productivity of building users.

6.3.1 BIM Use Cases for Reducing Process Times

The model-based quantity takeoff with BIM for tenders of (re)construction measures and FM services is significantly faster and more accurate than with traditional tendering methods. Through central model maintenance and direct integration of model-based quantity takeoffs into a tendering software, the bill of quantities can be created more efficiently.

In the context of space management or spatial equipment or furnishing planning, the power of imagination of users is considerably increased by BIM, and consequently decisions are made much faster and hardly need revision. Furthermore, inlets and clearances can be better checked with regard to relocation or equipment projects.

The effort for work preparation and scheduling of maintenance measures or the conversion, for example, of production facilities can be significantly reduced by using a BIM model. For many service requests, no on-site inspection is required in order to understand the installation situation or the installed type of defective equipments. Spare parts can already be ordered in advance because of the BIM model data. In this way, considerable preparation and travel times and thus costs are saved.

In order to carry out operational maintenance more efficiently and to counteract the shortage of skilled workers, BIM will play a significant role as an enabler in making these innovative technologies possible. So, Augmented Reality (AR) solutions providing remote assist make extensive use of BIM model data (Sects. 2.6 and 10.2.4). This will enable the usage of less qualified personnel being instructed digitally.

As part of the annual on-site inspections to identify the need for repair and renovation, BIM model information can support the estimation of costs for budget planning of measures, for example by providing the original construction costs. In addition, identified damage can be located on site also in the BIM model and then linked to measures with clear workflows.

The spatial representation of IoT measurement values in a BIM model allows for better orientation and easier detection of values that are outside range. Coupling a moisture monitoring on the construction site and in the building with the BIM model provides the exact location and enables faster and more targeted intervention in case of damage reports.

In the context of construction work for existing buildings or quality control of facility services, digital, centralized space- and asset-related approval reports can be created and edited supported by BIM CDE functionality. This includes both the location in the model using mobile input devices and the workflow-based processing of tasks and corresponding reports.

6.3.2 BIM Applications to Reduce External Costs

By working with a centralized and up-to-date data model and using the CDE functionality also in the operational phase, it is prevented that inventory data become inconsistent with each change in the inventory and eventually become outdated. In this way, areas and technical assets do not have to be re-entered or verified in an expensive way every three to five years before a new FM call for tenders. Here, not only external data capture costs, preparation costs of the call for tenders in terms of quantity takeoffs, but also startup costs on the FM service provider side are saved.

When reducing energy costs through rule-based optimization of HVAC and dynamic simulation in operations, BIM information can provide more accurate results through a better description of the physical building situation.

The BIM model-based thermal building simulation saves not only time when transferring the architecture into the simulation software, but also delivers more accurate results for later energy savings through demand-oriented design of the technical assets.

6.3.3 BIM Applications to Increase Productivity

As an example of how to increase the productivity of building users, daylight simulations can be improved based on BIM models. Here, not only time is saved when transferring the architecture into the simulation software, but the better illumination of the building based on the daylight simulation also provides more accurate insights into the later productivity increase of the users in the core business of the company.

In connection with *generative design,* for example, later production processes and their layout planning can be optimally coordinated with each other. Communication and traffic flows, logistics and production processes are optimized in this way and lead to an increase in productivity of the core processes of the company in the operational phase.

6.4 Evaluation of Benefits with the Balanced Scorecard

In Sects. 6.1, 6.2 and 6.3, the basic drivers for economic benefits and costs of using BIM in RE and FM and respective use cases for applying BIM efficiently in a construction project and the operational phase were explained in detail. However, the decision to actually implement BIM in an FM organization or a project for one or more of the explained BIM uses is not always easy. In particular, for the economic benefits described, the calculation of monetary values of savings by using BIM is not always possible. In particular, qualitative economic benefits, e.g. in the field of quality-related value creation (cf. Sect. 6.2.2), can only be insufficiently estimated with regard to the financial economic effects.

For strategic decisions that cannot be solely based on financial key performance indicators, the method of Balanced Scorecard (BSC) is generally accepted in practice. In general, however, the BSC method is initially associated with the introduction of methods and systems for performance measurement. Since the first publication of the BSC method by Robert S. Kaplan and David Norton in 1992 (Kaplan and Norton 1992), the BSC method has been very popular worldwide. In addition to the original objective, however, the BSC method can also be used very well to systematically determine and illustrate the economic benefits and feasibility of innovative, new methods and tools such as BIM.

6.4.1 Balanced Scorecard Method

This section first introduces the original BSC method, explains the basic principles and adapts it to the assessment of BIM use cases in the next step. The section concludes with the presentation of steps applying BSC-based evaluation in the context of BIM use cases.

The approach of BSC is based on the knowledge that management decisions are usually only made on the basis of financial indicators. However, qualitative aspects from both an internal and external perspective play at least an equally important role.

Kaplan and Norton therefore defined the two namesake basic principles of the BSC method.

Balanced

The term "balanced" expresses that, in addition to the financial perspective, other perspectives must also be taken into account for management decisions. In addition to the financial perspective (view of a company's shareholders), Kaplan and Norton define the customer perspective (sales and customer view), the process perspective (internal process organization) and the innovation or learning perspective (see Fig. 6.1). The BSC method now balances individual, sometimes contradictory, goals of the four perspectives through their target specifications. This allows the BSC to systematically include all aspects of a company strategy, even those whose goals cannot be represented by purely financial indicators.

Scorecard

The "scorecards" link the goals of the company strategy to specific indicators (Key Performance Indicator—KPI) for each of the four perspectives. This also includes the definition of targets and target values. Furthermore, for each goal, a reference is made to the measures required to achieve the goal. This answers the questions: "What do I have to do to achieve the goal?" and above all "Am I successful in implementing these measures?"

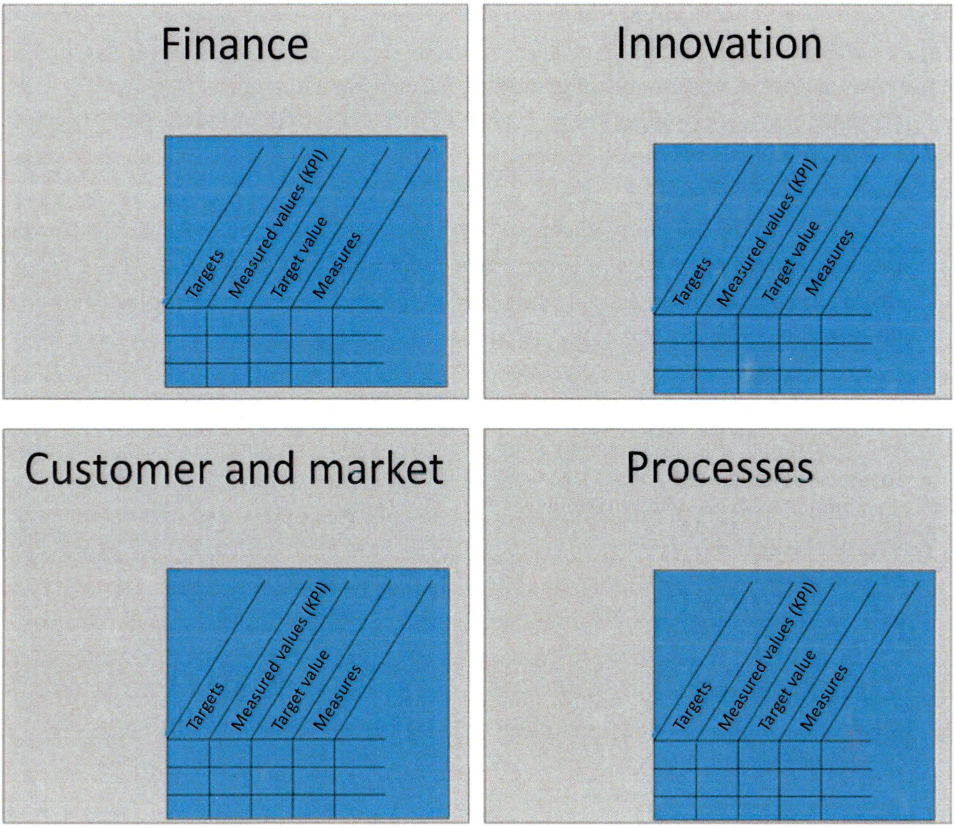

Fig. 6.1 BSC perspectives according to Kaplan and Norton

6.4.2 Application of the BSC Method for the Evaluation of BIM Use

In Sect. 6.2 and 6.3, the essential drivers for value creation through the use of BIM from the perspective of FM and the spectrum of benefits were presented. The problem of quantifying these benefits was discussed and, as the BSC approach suggests, it is also necessary to systematically include qualitative aspects in decision-making for the use of BIM. Consequently, a modification of the original BSC method is appropriate for this purpose.

If the BSC perspectives are used for decision making, no matter whether BIM should be used specifically for a project or even for the regular FM organization, the mentioned BIM benefits of BIM use cases replace the company's objectives in the BSC. The achievements of these benefits, whether in the BIM project or for the FM organization in general, is then the goal (target). Measuring the achievements of this goal thus enables to check whether the desired benefits have actually occurred.

The BSC approach will not only clarify the benefits in a structured way according to the recipients of the benefit (management, customer, organization, innovation), but also provides an instrument for checking the success of implementation.

BSC perspectives are as follows:

- *Financial perspective*
  This is where the monetary, financial benefits of BIM use are collected. Cost savings (e.g. quality costs) and new revenues expected from BIM use (e.g. additional services invoiced) are taken into account. It is expected that the key figures for measuring will be valued in monetary terms, e.g. in euros.
- *Customer perspective*
  This is where the benefits that provide added value to the client are listed. The client can be internal (e.g. other departments of the company) or external. For example, this includes benefits that using BIM generates for a FM service company for its (external) client. Benefits of this perspective can be both quantitative and qualitative.
- *Process perspective*
  This focuses on benefits that optimize internal business processes. The avoidance of rework due to errors in information transfer or -retrieval are typical examples. If the focus of the benefit is on the process perspective, e.g. the reduction of processing times is in focus. Resulting cost savings are only considered indirectly. However, if the focus is on cost savings, a shift of the BIM benefit to the financial perspective can be considered.
- *Innovation perspective*
  For many FM organizations, the BIM method represents a fundamentally new approach. This allows BIM pilot projects to gain first-hand experience and build specific BIM expertise. Typically, this benefit is considered in the innovation perspective. Cost savings are not considered yet. When updating the BIM BSC for the next projects, a corresponding adjustment in terms of the choice of the BSC perspective may be necessary.

6.4.3 Procedure for the Application of the BSC Method for the Evaluation of BIM Use

The application of BSC for evaluating BIM benefits and economic efficiency of BIM use can be divided into five steps, to be roughly evaluated in a workshop.

6.4.3.1 Collection of Benefits—Step 1

In the first step, the expected BIM benefits (benefit drivers) are only collected. For this purpose, the presentation of the benefits in Sects. 3.3, 6.2 and 6.3 can be used. The compilation of BIM use cases (BIM uses) of the *BIM Project Execution Planning Guide* (Kreider and Messner 2013) also provides a very good overview of typical BIM benefits throughout the entire building lifecycle. The authors describe different BIM use cases structured according to purpose, typical features of the BIM use and the BIM competencies required for implementation. In addition, assistance is provided in selecting suitable BIM use cases. In the field of infrastructure construction projects, part 6 of the initiative BIM4INFRA2020 resulted in a compilation of most important BIM use cases, which also provide information on potential benefits, the effort required for implementation as well as relevant data, models and formats (Borrmann et al. 2019b).

6.4.3.2 Operationalization of the Benefits—Step 2

In a next step, the selected BIM benefits are now described in a structured manner for the specific project. The results of step 1 can be used as a template for BIM use cases.

The following information is specified for each BIM use case:

- *Objective* of the BIM use case
  Which benefit is expected to be achieved with the use of BIM for the project?
- *BIM roles/BIM competencies*
  Which roles and project partners are involved in this BIM use case and which BIM competencies are required for its implementation?
- *Information objects*
  Which data, possibly also which models (e.g. existing models, special discipline models) and formats are required for this BIM use case?
- *Notes/opportunities and risks in implementation*
  Which challenges need to be considered when implementing the BIM use case?

6.4.3.3 Assignment of Benefits to a BSC Perspective—Step 3

The third step now defines from which BSC perspective the benefits of the BIM use case are to be considered for the decision as to whether and to what extent the BIM method is to be applied.

For the benefit of *optimizing technical spaces* explained in Sect. 6.2.3, an assignment to two of the BSC perspectives might make sense:

- Assignment to the *financial perspective*
  This perspective would be chosen if the company itself benefits from the financial savings effect by reducing the technical spaces in the planning phase of a construction project later in the operational phase. For example, additional space gained through optimization can be used for leasing to generate additional revenue.

- Assignment to the *process perspective*
  The process perspective would be chosen for this benefit if the maintenance processes were improved by optimizing technical space. As explained in Sect. 6.2.3, an undersizing of technical spaces leads to missing working space for maintenance and reduced accessibility of assets. The optimization of working space around a technical asset therefore leads to lower maintenance costs and an improved process quality for operations in the end.

The assignment of a benefit can also be made to multiple perspectives, with a priority usually being given to one perspective.

6.4.3.4 Assigning Indicators to Measure the Achievement of Objectives—Step 4

One of the great advantages of the BSC method is linking objectives (benefits) with the addressees of a perspective (e.g. stakeholders/shareholders, customers, internal organization), as well as linking objectives with key performance indicators (KPI's) to measure the achievement of objectives. In this case, qualitative indicators are regularly chosen as KP'Is for benefits. Only in the financial perspective quantitative key figures are generally used.

In the example chosen in Step 3, the benefit "Optimization of technical spaces" would be measured in the case of assignment to the financial perspective with the expected additional revenues as a key figure. The size could be given absolutely in euros or relatively, in a percentage increase in total revenues from rental.

If, on the other hand, the process perspective is chosen in Step 3, an expected reduction in processing times for maintenance activities could be chosen as a key figure. However, qualitative effects such as the improvement of process quality are also conceivable.

6.4.3.5 Determining the Measures to Achieve Objectives—Step 5

In the last step, the measures for implementing BIM use cases are specified. For example, information requirements are translated into EIR and required processes and Data Drops are described in BEP or IDM (see Sect. 3.4 and Chap. 7).

6.5 Summary

The continuous application of BIM also in the operational phase of buildings often fails today due to the skepticism of FM and/or owner organizations about the economic benefits of BIM, especially—but not only—for existing buildings.

This chapter first explains the basic drivers for the economic efficiency and ROI of BIM uses. Next, typical benefits of BIM in the construction phase are discussed, which have also an impact on value creation in operations. This leads to using BIM as "digital prototyping" and "digital lifecycle management" in the design and planning phase. The benefits resulting from this process-related consideration are supplemented by the quality-related value creation in the construction phase, whereby the advantages of BIM are explained by a low-threshold, early application of simulation methods. In the design and planning phase, innovations can still be taken into account in a timely manner by new generations of technical assets, as the effects on planning can be estimated and calculated with significantly less effort by exchanging individual, more energy-efficient components. A resource-related consideration of value creation by the use of BIM from the perspective of space and energy efficiency concludes the construction phase with practical experiences in the optimization of technical areas by a BIM-based planning.

In the operational phase, benefits through the reduction of process cycle times of important processes, such as maintenance processes, are first presented using BIM. For example, it is explained how BIM models act as "enablers" for the use of innovative technologies such as augmented or mixed reality. The use of these technologies not only results in efficiency gains in terms of processing speed, but also decisions on, for example, the exchange or repair of technical components are made faster and with less mistakes. The benefits of BIM are of high imporatance due to todays shortage of skilled workers, since it also enables approaches to provide remote support of on-site personnel. Finally, examples of reducing external costs, such as avoiding costs of on-site data capture when exchanging service companies, are explained. Furthermore, benefits are explained by the use of BIM to increase the productivity of building users by better daylighting.

The last section of this chapter proposes a new approach to the evaluation of BIM benefits in the construction and operational phase. The well-established approach of the balanced scorecard (BSC) for corporate strategic decisions that are not based solely on financial indicators is transferred to the evaluation of BIM benefits. Here too, it is often not possible to justify the use of BIM solely by financial savings. Rather, a decision must often be taken depending on qualitative benefits of the BIM uses for customers, one's own process organization or one's own innovation capability. The section ends with presenting a five-step approach for the practical application of BSC-based evaluation in the context of BIM use cases.

References

Borrmann A, Elixmann R, Eschenbruch K, Forster C, Hausknecht K, Hecker D, Hochmuth M, Klempin C, Kluge M, König M, Liebich T, Schöferhoff G, Schmidt I, Trzechiak M, Tulke J, Vilgertshofer S, Wagner B (2019b) Steckbriefe der wichtigsten BIM-Anwendungsfälle. Publikationen BIM4INFRA 2020, Teil 6

Kaplan R, Norton D (1992) The Balanced Scorecard—Measures That Drive Performance. In: Harvard Business Review, Jan–Feb 1992

Kreider R, Messner J (2013) The Use of BIM. Classifying and Selecting BIM Uses. PennState University College of Engineering. https://pennstateoffice365-my.sharepoint.com/:b:/g/personal/jim101_psu_edu/EYm_wQdsDn5MvcFwDbrg-SsB7LGn7iP5_WazMXwFdVFDZQ?e=iod4JD (retrieved: 18.11.2021)

May M (Ed.) (2018a) CAFM-Handbuch—Digitalisierung im Facility Management erfolgreich einsetzen. 4th edn., Springer Vieweg, Wiesbaden, 2018, 713 S

NN (2016a) GEFMA Richtlinie 460: Wirtschaftlichkeit von CAFM-Systemen, May 2016, 27 p

NN (2019h) Energieeffizienzstrategie 2050. Bundesministerium für Wirtschaft und Energie. https://www.bmwi.de/Redaktion/DE/Publikationen/Energie/energieeffiezienzstrategie-2050.html (retrieved: 18.11.2021)

BIM Implementation in RE and FM Organizations

7

Maik Schlundt, Thomas Bender, Michael Härtig, Erik Jaspers and Marko Opić

7.1 Stakeholders in BIM4FM Projects

Stakeholders are people, groups or organizations that can be affected by a BIM project. The ambitions and goals of stakeholders often complement each other, but can also be different and, in the worst case, even contradictory. Not everyone has the same influence on the development of the BIM method in the respective project.

M. Schlundt (✉)
DKB Service GmbH, Berlin, Germany
e-mail: maik.schlundt@dkb-service.de

T. Bender
pit – cup GmbH, Heidelberg, Germany
e-mail: thomas.bender@pit.de

M. Härtig
N+P Informationssysteme GmbH, Meerane, Germany
e-mail: michael.haertig@nupis.de

E. Jaspers
Planon B.V., Nijmegen, The Netherlands
e-mail: erik.jaspers@planonsoftware.com

M. Opić
Alpha IC GmbH, Nürnberg, Germany
e-mail: m.opic@alpha-ic.com

M. May et al. (eds.), *BIM in Property Management*,
https://doi.org/10.1007/978-3-658-40830-5_7

7.1.1 Stakeholders During the Building Lifecycle

The consideration of the lifecycle phases according to GEFMA guideline 100 (NN 2004a) of a building opens up the view to many stakeholders of a BIM project (see Fig. 7.1).

First of all, it is noticeable that the supposed "main protagonists" of BIM, architects, planners and builders, compared to the entire lifecycle of a building, only have a small stakeholder share concerning the duration of their contribution in the lifecycle. In particular, the distinction between data generation and data use clarify the role of BIM in the building lifecycle phases after construction. It should be noted that the real duration of the different phases are not shown to scale in Fig. 7.1 —the phases of building operations are known to be much longer than the construction phases.

7.1.2 Data Creators

The main actors in BIM projects are *architects and planners.* Their collaboration on the BIM model is the key to achieving efficiency and benefits through BIM with regard to decision-making, time and cost planning or construction management processes. With the help of BIM, they ensure a high degree of planning speed and reliability. They are focused on the successful completion of the building. However, the implementation of

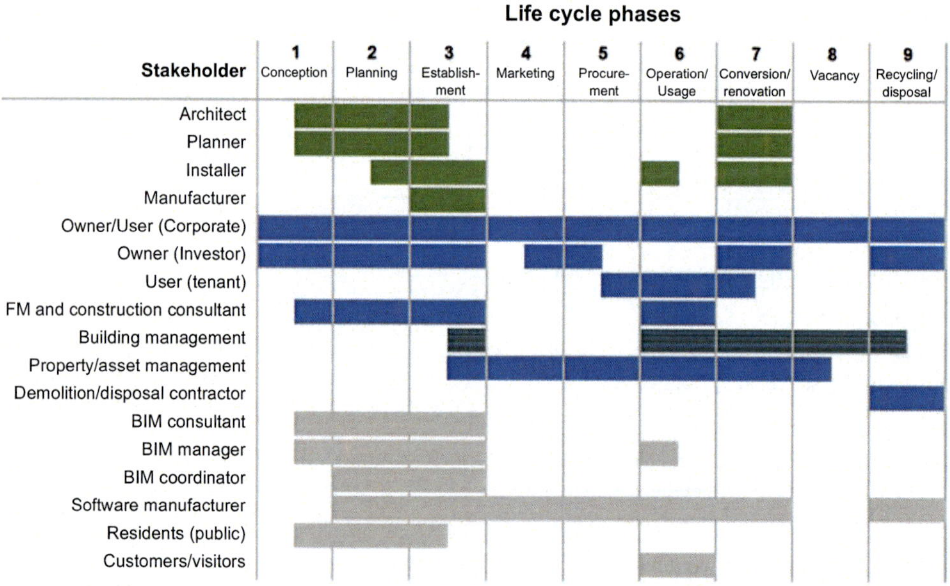

Fig. 7.1 Periods of occurrence of different stakeholders in the BIM process (data generation—green, data use—blue, consulting/support/other—gray).

BIM requirements from building operations are often perceived as resulting in additional effort and therefore are sometimes seen to be contrary to the economic success of the project from the perspective of the construction team.

This is particularly true for the *contractors*, whose activities are closer to the operational phase and who are therefore more intensively involved with the corresponding requirements (e.g. for data exchange and documentation).

As indirectly involved, the *manufacturers* and suppliers of building and technical components currently play an underestimated role. In addition to the functional and material quality of delivered components also the provision of comprehensive digital data (cf. Sect. 6.3) represents an added value which is not yet recognized as important from a whole life perspective. This has particular consequences for the interface to FM, for which corresponding information is mandatory.

7.1.3 Data Users

Among the data-using stakeholders, the *developer*s are is in first place. Their interest to achieve the best possible quality using BIM both in the planning and construction process and regarding the building documentation and facility operations is great, but here too cost aspects are often in the foreground. Two types of development are to be distinguished: Those who construct the building for later self-use (as Corporate Real Estate) typically give greater importance to benefits in the operating phase than a classical investor, who only holds the object for a very limited time according to the business model and therefore does not consider the entire lifecycle, but rather focusses on the documentation in the transaction phase. However, both benefit from an accurate and above all quickly available as-built documentation.

As the *tenant* of the object, the building *user* is rather an indirect stakeholder of BIM. They benefit from efficient building operations and possibly a higher technological quality of their rental space. In this respect, they can later integrate their own internal services (e.g. based on IoT, see Sect. 2.7) into the BIM model.

The *FM* and other *building consultants* are essential to provide inputs about operational needs during the planning and construction process. They usually do not generate data within the BIM model, but rather use it to support the tendering of FM services by means of data export of technical assets and floor plans or to carry out simulations of material flows or other physical simulations making use of the geometric and material data in the model (cf. Sect. 2.11). In a BIM project, an FM consultant may have another role: as a specialist regarding the operational phase, they can make a significant contribution to the requirements definition of the BIM model documented in EIR or AIR (cf. Sect. 3.4).

Both as a user and producer of BIM data, FM is one of the most important stakeholders of a BIM project. By gaining knowledge of the BIM process and timely involvement (during definition of information requirements), later object operations can be optimally aligned with the use of the BIM model and designed efficiently with regard to documentation processes. Only under this condition is further data maintenance of a digital twin (cf. Sect. 4.1) conceivable. Of course, BIM-based building operation poses high demands on data maintenance by architects and planners, and especially by the construction companies, and thus generates related costs.

Other stakeholders during operation are property and *asset management*. Their need for data from BIM is strongly dependent on the relationship to building management, but will rather focus on infrastructure data, especially space and inventory information.

To emphasize the relevance of BIM throughout the entire building lifecycle, the last phase, disposal, should also be mentioned. Here, *demolition, recycling* and *disposal companies* can obtain valuable data from the building structure to information on used materials to quantity information in order to ensure safe dismantling, the most comprehensive reuse or recycling, and unavoidable disposal.

7.1.4 Consultants and Supporters

In the planning and development phase of a construction project, various consulting and supporting stakeholders play a role, but neither generate nor use data. Thus, *BIM managers* and BIM coordinators share the common goal of efficient BIM project implementation based on EIR-derived BEP. However, as representatives of various construction participants, the ideas of achieving the goal can vary greatly. An externally hired *BIM consultant,* who can also act as a mediator, is often a helpful supplement to the project team.

The software vendors of products used during the planning and construction phase as well as during later operations have an influence on successfully using BIM. In addition to vendors of BIM software, the suppliers of CDE platforms, simulation software, CAFM and many other systems that enable digital planning, construction and operation processes based on BIM data are to be mentioned. The interfaces required for this are increasingly standardized, which is only possible through the cooperation of all market participants and the commitment of organizations and associations (see Annex 2).

7.1.5 Other

Two essential participants or those affected by construction projects should not be forgotten. In particular, with buildings of public interest, it is often necessary to comprehensively inform *residents* or the *public* interested for other reasons. Quick and reliable answers to specific questions are to be presented convincingly, which is more likely to succeed with an integral and graphically high-quality 3D building model than with notice boards with views and sections.

Other stakeholders can be *customers* and *visitors* of the future building users. They benefit from many technical solutions based on the data of BIM models, such as indoor navigation. The satisfaction of customers and visitors is a driver that can have an impact on the future development of BIM via the building users.

7.2 Approach in a BIM Project

The number of construction projects that are being implemented using BIM is rapidly increasing. In most BIM projects, however, the focus is on the planning and construction phases. Subsequent building operations is still often rarely taken into account in BIM projects.

In order to generate added value for building operations from a BIM project, it is essential to consider the requirements from FM at the beginning of the project and to implement them consistently. In addition, the required BIM roles and the associated IT environment are to be established (see Fig. 7.2). Here, the BIM data manager plays a fundamental role (see Sect. 5.4).

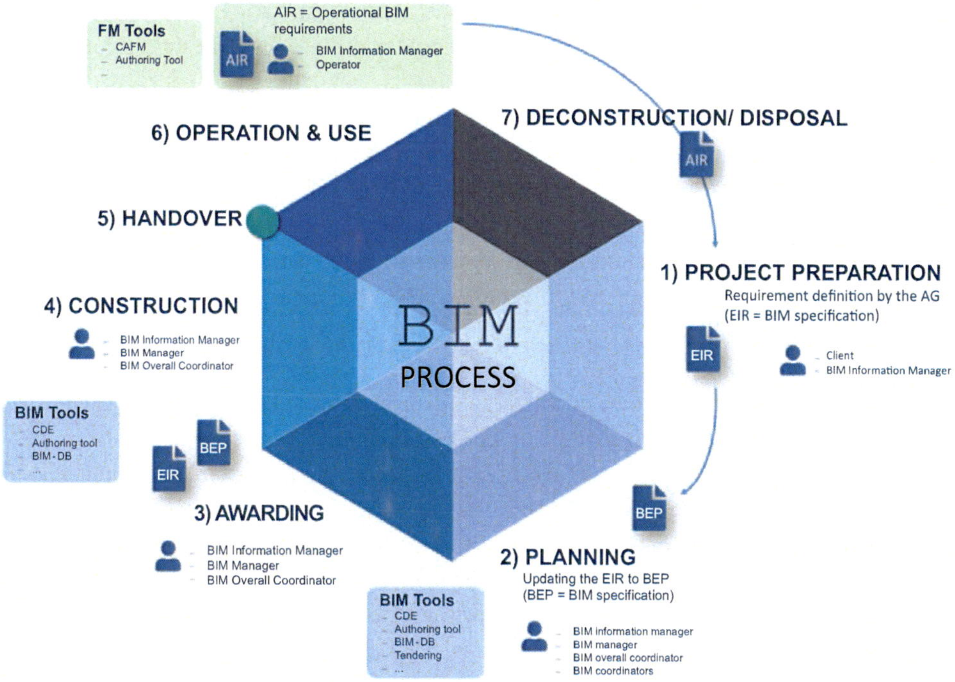

Fig. 7.2 Schematic representation of the BIM reference process

From an FM perspective, the following aspects must be taken into account and anchored in a BIM project:

- Describing requirements from FM for a BIM project (Asset Information Requirements—AIR),
- Integrating the AIR into the BIM project documents (Employer Information Requirements—EIR),
- Implementing the roles required from an FM perspective in the project,
- Establishing BIM tools relevant from an FM perspective in the project, such as authoring tools, BIM DB and CDE.

7.2.1 Requirements from Facility Management

The requirements from FM for BIM or for a BIM project should already be taken into account when the constructor's goal and strategy are defined. The FM requirements are specified by describing the OIR and AIR.

The OIR summarize key information needs for reporting and managing organizations day to day business. The AIR document summarizes data and information requirements for the operational phase of a property portfolio of an organization. It specifies the information needs for all stakeholders to plan, control and manage real estate throughout its lifecycle (NN 2019e).

Depending on how the future building operations is to be designed (what processes are established, which systems are used, which data and documents are required for this?), accurate information requirements can be derived. The requirements are created by the owner in collaboration with the building operator and/or an FM consultant. They are usually defined organization-wide and can be used for various BIM projects.

In this context, the AIR with the perspective of building operations provide the following information:

- General definition of a consistent, logical identification of buildings, spaces and structural and technical objects using a unified asset identification system,
- Definition of properties (attributes) of objects according to classification standards for FM such as CAFM-Connect,
- Data drops—which content is delivered, when and by whom—defining the Level of Development (LOD)/Level of Information (LOI),
- Data formats (e.g. IFC, COBie) for lossless data exchange with CAFM and other IT systems.

The aim of the asset information requirements is to describe the requirements of FM for a BIM project and thus all deliveries of the project participants to the as-built documentation in a binding manner. The result is a valid and structured data basis which can be transferred or integrated into building operations without any problems after the end of the project.

7.2.2 BIM Project Documents

There are different types of documents in a BIM project that need to be created as a basis for a regulated project process. These documents are a significant part of the contracts with each project participant (e.g. architects, MEP planners, general contractor).

In addition to the AIR, these are essentially the employer information requirements and the BIM execution plan (cf. Fig. 7.3).

7.2.2.1 Employer Information Requirements (EIR)

In the employer information requirements the client (building owner) defines his requirements for the use of BIM in the project and derives the requirements for the creation of information. The EIR essentially provide information about which information is required and to what degree of detail. They thus describe the "WHAT" that is to be delivered from the contractor to the employer.

The EIR are part of the tender documents and are intended to inform the bidders comprehensively about requirements and information needs of the client so that they can assess and prove their competence in terms of BIM. The EIR are project-specific and must always be adapted to the requirements of the respective BIM project.

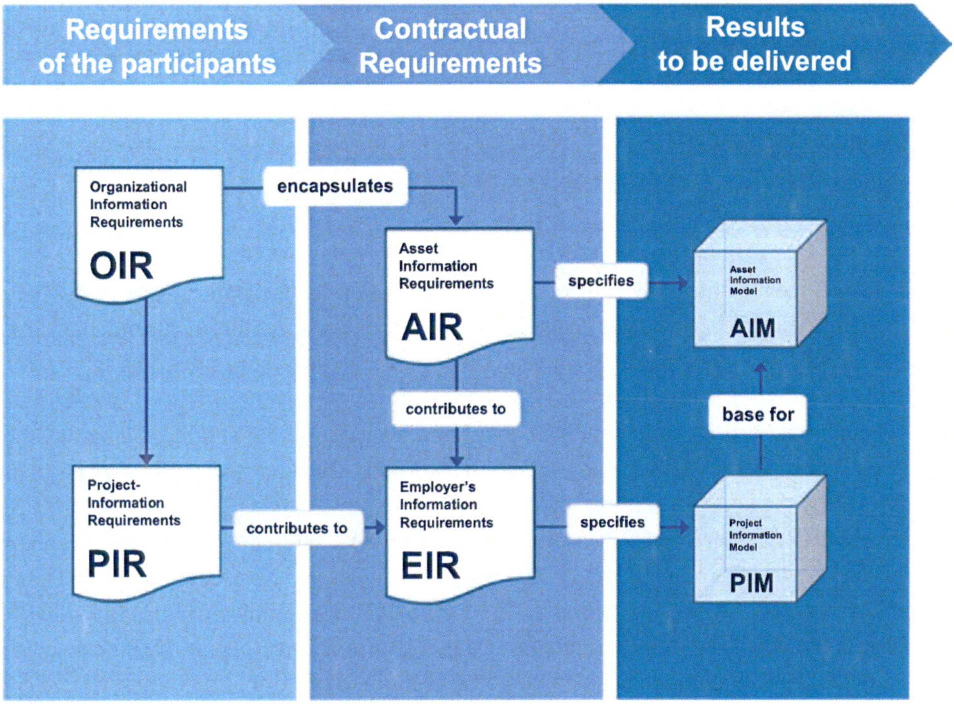

Fig. 7.3 BIM documents

The following topic areas must be described in detail in the EIR (NN 2015d):

- Technical details (software platform, LOD/LOI definition, etc.),
- Management details (description of the management process to be set up in connection with a BIM project),
- Commercial details (delivery services, handover dates).

The requirements of FM OIR and AIR are to be integrated into the EIR or reference is to be made to them.

The BIM4INFRA2020 working group is commissioned by the German Federal Ministry of Transport and Digital Infrastructure (BMVI) to provide scientific support in connection with the implementation of Building Information Modeling. A total of 10 guidelines have been issued by the working group as recommendations for dealing with BIM. In Part 2: Guidelines and template for employer's information requirements (EIR), the working group describes the framework for EIR as follows:

Table of EIR contents (NN 2019f):

1. BIM Applications
2. Digital Foundations Provided
3. Digital Deliverables
4. Organization and Roles
5. Collaboration Strategy
6. Delivery Dates
7. Quality Assurance
8. Model Structure and Model Content
9. Technologies

Which services in the BIM project are to be provided by the participants is described in the EIR via BIM use cases. From these use cases arise requirements for the digital deliverables to be created. An overview of possible use cases is given in Fig. 7.4 (NN 2019d).

The contractor has to create, check and hand over digital deliverables as part of their services. The digital deliverables are to be created, checked and handed over to the employer by the contractor in the BIM project.

In the EIR, the deliverables are also described with reference to the service phases according to the German HOAI (Official Scale of Fees for Services by Architects and Engineers) and with the indication of the LOD. The deliverables are usually files that are to be handed over to the employer at the end of a service phase. These can be, for example, 3D building models, calculations, alphanumeric information or descriptive documents. The level of development (LOG, LOI) is to be described in detail in the EIR.

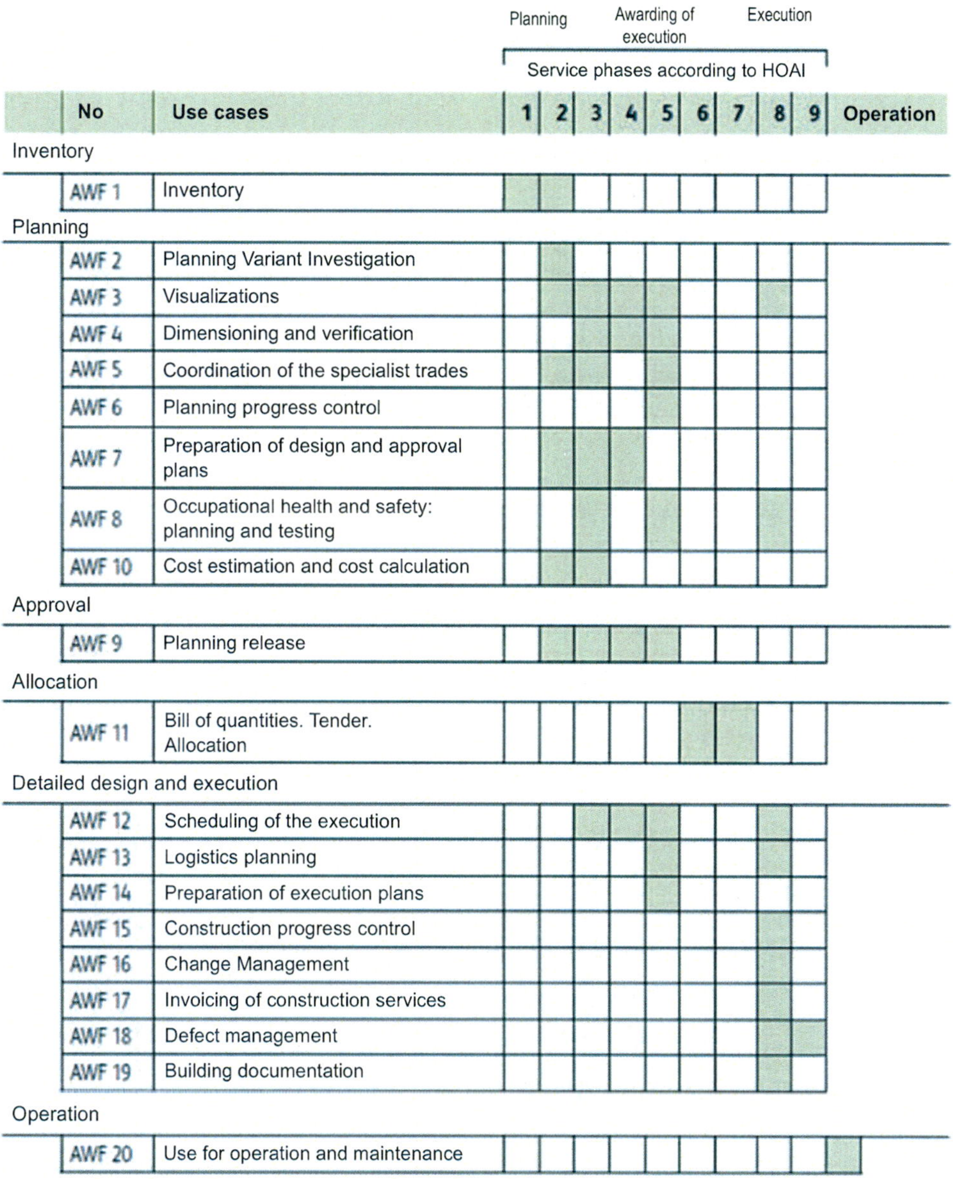

Fig. 7.4 BIM use cases

7.2.2.2 BIM Execution Plan (BEP)

Often before contracts are awarded a pre-contract BEP is prepared to make sure that the contractor has understood clearly the clients EIR. After awarding, the project participants who have been determined at this time create the BIM execution plan (BEP), in which it is specified how the employer information requirements are to be met. While the "WHAT" is described in the EIR, the "HOW" is described in the BEP.

The BEP is created and updated on the basis of the EIR and hence forms the project-specific backbone of a BIM project with regard to the creation, transfer and management of data and information. It answers among others (May 2018a):

- How the roles and responsibilities are distributed,
- Which technology is used,
- How often and when planning meetings are held and
- Which parts of the planning are to be modeled and when and to what degree.

The requirements defined in advance from FM OIR and AIR are to be integrated into the EIR and the BEP. Both EIR and BEP are basic contract documents in a BIM project and are then to be implemented mandatory for the project participants.

7.3 Common Data Environment (CDE)

A central, shared data environment CDE (Common Data Environment) is the focal point in a BIM project for

- Collection,
- Management and
- Distribution

of all elements of the BIM information model. In this context, the term information model refers to the combination of geometric information (model), structured semantic (alphanumeric) data and documentation. The CDE consists of processes, conventions, rules and supporting technologies to fulfill the above-mentioned tasks (see in detail in Sect. 4.3).

7.4 Roles in the BIM Project

In comparison to a traditional construction project, a BIM project requires new roles with new tasks, responsibilities and accountability. Basically, these are:

- BIM information manager,
- BIM manager,
- BIM project coordinator,
- BIM disciplinary coordinator(s) architect, specialist planner.

Depending on project size, project organization and procurement strategy, the roles can be taken over by different project participants (e.g. employer, architect), possibly also in a personal union. If necessary, some roles such as the BIM project coordinator are not used.

7.4.1 BIM Information Manager

The BIM information manager is the point of contact for BIM content from the perspective of the client. They define information needs and model requirements of the client and ensure implementation in the project. The BIM information manager is responsible for creating the EIR. Analogous to the BIM data manager (cf. Sect. 5.4), who is responsible for data quality in the operational phase, the BIM information manager represents, inter alia, the perspective of operations in the planning and construction phases.

7.4.2 BIM Manager

The BIM manager is responsible for setting up the BIM project and organizing management processes required for this. They ensure consistent handling of the model and the documents derived from the model. Therefore, they are responsible for creating the BIM execution plan (BEP).

7.4.3 BIM Project Coordinator

The BIM project coordinator supports cooperation and coordination of model information and monitors the model quality in accordance with the project guidelines and requirements.

7.4.4 BIM Coordinator

The respective BIM coordinator of the different disciplines has extensive knowledge of the BIM method and coordinates internal requirements with the project demands. They ensure model consistency among the disciplines and is responsible for quality assurance of all data before they are released to other project participants.

The interests of FM in a BIM project are protected by the BIM information manager and the BIM manager and forwarded to the BIM project coordinator and to the disciplinary BIM coordinators. If necessary, these roles need to draw on the expertise of FM in order to translate requirements from FM correctly and finally to monitor their implementation up to the handover of data to building operations.

If these tasks are not carried out with the required level of detail and consistency, a smooth data transfer to building operations cannot be guaranteed.

7.5 Application Scenarios

7.5.1 Commissioning and FM Handover Phase

If BIM is divided into the phases planning and construction as well as operations, the short but essential commissioning and handover period of the building to the client is often neglected and overlooked. But this transitional phase is most important for the application scenarios in building operations.

With the completion of the classical planning phases, the transition to construction of the building begins. At this transfer point, the BIM model already contains most of the data relevant for later operations. These were ideally already defined in an operating concept during the early design planning and further elaborated in the subsequent planning process. A BIM-based tendering of FM services now allows the early involvement of the future operatoring service company. When this is done, the designated operating company can ensure ideal conditions for efficient building operations and accompany the critical commissioning phase.

The provision of BIM models to bidders of FM services ensures easily accessible and comprehensive information resulting in more accurate calculations of required FM services. Hence, possibly time-consuming and at the time of construction usually only limited on-site visits can be dispensed with. The export of building data simplifies quantity takeoffs for the bill of materials and services as well as price lists for calculations. If the FM service provider has done the calculations based on BIM data, the synchronization of services and quantities during implemenation is significantly easier, since it can be done ideally by an automatic update of the BIM data.

For FM service providers, this option creates a quite interesting field of activity, which is currently getting little attention in the service industry. Before a BIM model can be used as a digital twin (see Sect. 4.1), the deviations between execution and planning that inevitably occur in every construction process must be corrected. Only then does it correspond to the actual construction state (as-built). When this service is offered by an FM service provider, many synergies could be achieved, such as gaining best possible knowledge of the building or fulfilling the employer's request for a neutral review of the BIM model.

If the BIM model is not to be used as a digital twin, the facility manager can transfer the information captured during planning and execution as an initial data set into the CAFM system (see Sects. 3.2 and 5.3).

7.5.2 Operational Phase

In order to be able to benefit from BIM data in the operational phase, IT support of FM processes is essential. As a rule, operators use CAFM software to manage their facilities.

A continuous data synchronization between BIM and CAFM is essential. The data of the BIM model mainly provide architectural and technical descriptions of the building and its assets, which are transferred to CAFM and supplemented with all FM process-relevant data. During further operations, data changes can affect the other system (see Fig. 7.5). The degree of integration of both systems is flexible.

The following section shows various application scenarios that need interaction of BIM and CAFM.

Scenario	• Data/action
Tender for FM services	• Export of (alphanumeric) bulk data and possibly graphical data from CAFM
	• Digital walk-through in the BIM model
Create the CAFM data model	• Import 3D data from BIM model
	• Import alphanumeric data and documents
Enriching the CAFM data	• After planning of conversions in BIM, data in CAFM are updated or supplemented
Enriching the BIM model	• After maintenance measures, for example, new asset components are transferred to BIM
Visualization of data in the BIM model	• Color coding of data in the model, visualization of relocation variants • Serves as a basis for decision-making

Fig. 7.5 BIM scenarios in FM

7.5.3 Maintenance

The quality of maintenance depends to a large extent on precise knowledge of all the building components and technical assets. This starts with detailed information on system components (e.g. manufacturer/product, type, performance data) and extends to their functional relationship (supply area) up to the exact location in the building.

The description of technical assets already created in the planning phase in the BIM model and their supplement with additional data by installation companies give the persons responsible for maintenance all the information they need for targeted service provision. In addition, the current installation environment that can be traced at any time in the model ensures best preparation, e.g. for a repair. The visualization of the supply/service area of technical assetss quickly shows which building parts would be affected by a shutdown.

The BIM model also enables information technologies that have so far hardly been economically implemented in buildings. Technical installations are automatically detected by the use of augmented reality glasses (cf. Sect. 2.6) and relevant technical information on the respective system component is displayed in the field of view of the person carrying out the work. Indoor navigation helps people to find their way to the job site safely even in large and complex buildings.

Of course, it is of paramount importance that the data is updated accordingly after a maintenance measure has been completed.

7.5.4 Move Management

CAFM products already offer very good support for planning and visualizing move variants based on 2D plans. However, the information density of a BIM model is much higher compared with 2D plans. For example, data on MEP installations can be accessed without time-consuming on-site visits and requirements or potentials for conversions can be more easily recognized. In addition, the view of the entire building instead of individual floors allows a deeper insight into functional and communicative relationships between departments and areas of use.

7.5.5 Smart Building

Sensors and actuators as part of the Internet of Things (IoT) are important components of an increasingly digital future of real estates. As components located in the BIM model, they provide the basis to control various FM processes. For example, sensors can be used to determine air quality in order to visualize critical areas in the model and to control the corresponding technical assets. Occupancy sensors are particularly effective in supporting the processes of demand-driven cleaning and space utilization.

7.6 Summary

This chapter deals with how BIM can be successfully implemented in real estate and facility management. For this purpose, the stakeholders are introduced with respect to data management. While BIM serves as primary tool for planners and architects and therefore is a tool with great impact on decisions in the design and planning phase, questions about subsequent operations are usually only asked very late in the project. However, FM uses BIM data over the longest lifecycle period and may also be responsible for the maintenance of models and, in particular, a digital twin. But many other stakeholders in construction and operations also have direct or indirect influence on a BIM project. The early identification and consideration of as many requirements as possible can contribute significantly to the success of the project.

In addition, the steps in a BIM project is presented with requirements for project documentation, the roles involved and their tasks. Only the consideration of the specific requirements of RE and FM ensures success of the BIM method.

BIM can already provide important data for FM during construction and commissioning. In order to use BIM successfully in all FM processes, the execution of a BIM model as a digital twin is helpful. For further digitalization of real estate, the BIM model plays a significant role.

References

May M (Ed.) (2018a) CAFM-Handbuch—Digitalisierung im Facility Management erfolgreich einsetzen. 4. edn, Springer Vieweg, Wiesbaden, 2018, 713 p

NN (2004a) GEFMA Richtlinie 100-1: Facility Management—Grundlagen, July 2004, 21 p

NN (2015d) Employers Information Requirements—Structure of an EIR. https://toolkit.thenbs.com/articles/employers-information-requirements (retrieved: 10.11.2021)

NN (2019d) BIM4Infra2020, Teil 1—Grundlagen und BIM-Gesamtprozess, April 2019

NN (2019e) Bauen digital Schweiz, LIM Liegenschafts-Informationsmodell/IMB Informationsmodell Bewirtschaftung, Arbeitsdokument, August 2019

NN (2019f) BIM4Infra2020, Teil 2—Leitfaden und Muster für Auftraggeber Informationsanforderungen (AIA), Abschnitt II Muster AIA, April 2019

8 BIM in FM Applications

Michael May, Nancy Bock, Michael Härtig, Joachim Hohmann, Markus Krämer, Bernd Limberger and Marko Opić

8.1 CAFM System Support

The application of BIM is still mainly used in the building lifecycle phases of planning and construction (see Fig. 7.1). Here the utilization of technologies to improve efficiency of operations leads to quick benefits and the return on investment (ROI) is quite easy to determine and assign (see Sect. 6.4).

Of course, cost and benefit considerations can also be made for the operation of real estate (see Sect. 6.3). However, there is a long period of time between the decision to use BIM in FM, which is best taken in the design phase, and achieving the related benefits (sometimes after a few years of operation). During this period decision-makers, cost-

M. May (✉)
Deutscher Verband für Facility Management (GEFMA), Bonn, Germany
e-mail: michael.may@gefma.de

N. Bock
BuildingMinds GmbH, Berlin, Germany
e-mail: nancy.bock@buildingminds.com

M. Härtig
N+P Informationssysteme GmbH, Meerane, Germany
e-mail: michael.haertig@nupis.de

J. Hohmann
Technische Universität Kaiserslautern, Kaiserslautern, Germany
e-mail: joachim.hohmann@bauing.uni-kl.de

M. Krämer
Hochschule für Technik und Wirtschaft Berlin, Berlin, Germany
e-mail: markus.kraemer@htw-berlin.de

M. May et al. (eds.), *BIM in Property Management*,
https://doi.org/10.1007/978-3-658-40830-5_8

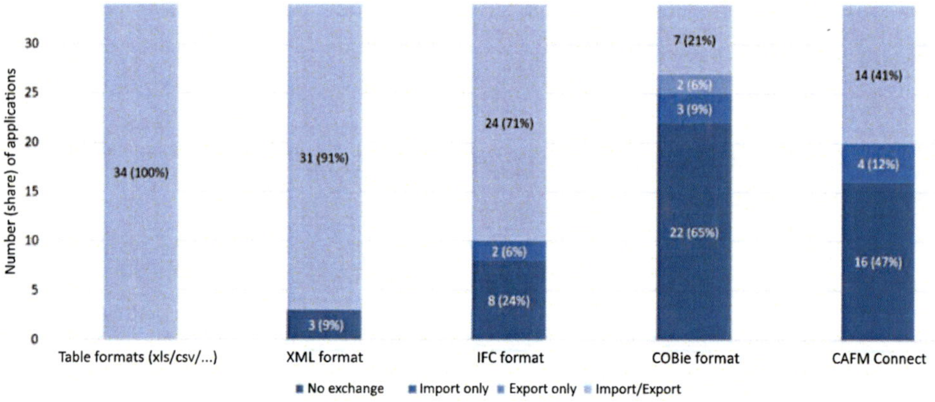

Fig. 8.1 Support for indirect access to BIM models via interface formats (N = 34, multiple answers possible)

bearers and ultimately the ROI beneficiaries are in their various roles (owners, project developers, investors, planners, contractors, users, service providers, etc.) and open collaboration is often hard to achieve.

However, technologically, the use of the BIM method for real estate operations is mature enough that more and more CAFM providers are enabling the connection of CAFM and BIM via special interfaces (see Fig. 8.1)

Necessary interface formats such as IFC, COBie or XML, as well as direct connections to tools in BIM environments, such as Autodesk Revit or other CAD/BIM solutions, have already been integrated into many CAFM products. All software vendors participating in the "CAFM Software Market Overview" (NN 2021b, see also NN 2022b) state that they can read/write data from/into a BIM model via interfaces (Fig. 8.1).

The corresponding abilities of CAFM software for BIM data import and export are also referred to, which are checked as part of the CAFM certification according to GEFMA guideline 444 (see NN 2020a and Sect. 8.2).

At the level of data exchange, there are already diverse possibilities for communication between systems. However, BIM and CAFM only represent parallel data and process environments that are in no way integrated. However, FM processes cannot be mapped in BIM directly, only the metaprocesses for maintaining BIM data.

B. Limberger
SAP Deutschland SE & Co. KG, Walldorf, Germany
e-mail: bernd.limberger@sap.com

M. Opić
Alpha IC GmbH, Nürnberg, Germany
e-mail: m.opic@alpha-ic.com

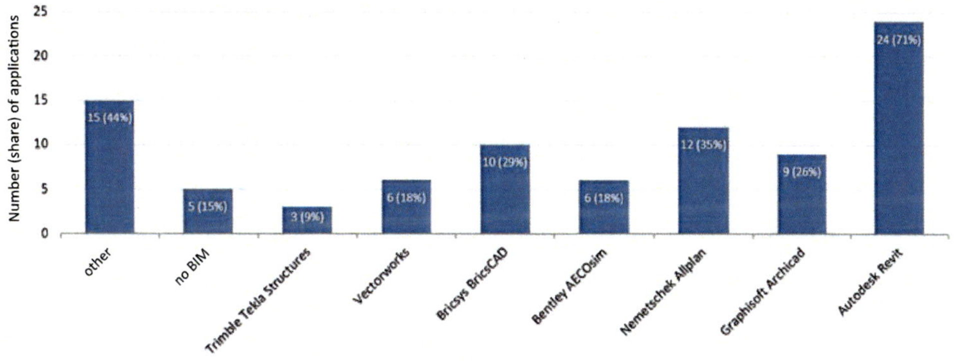

Fig. 8.2 Support of access to BIM models via direct interfaces to authoring tools (N = 34, multiple answers possible)

As an example of application from the planning and construction phase, the preparation of FM-relevant data for the utilization in FM should be mentioned. If planners and builders are commissioned to capture operation-relevant data, these data can be transferred step by step from the BIM model to a CAFM software and, for example, the early procurement of required FM services can be controlled. If the FM service provider is already fixed (because it is another building in a larger portfolio or the builder also provides FM services), they can specify the data structures using their CAFM software, transfer them into the BIM model and, ensure the data quality during the project.

In addition to this *data exchange level*, many software vendors also support direct access to BIM authoring tools. Fig. 8.2 (NN 2022b) shows that approximately 70% of CAFM products can interact directly with the Autodesk Revit authoring tool. This enables direct communication between BIM and CAFM on the *process level.*

However, the focus of most products is currently on the visual representation of buildings as a 3D model within the CAFM software, often supplemented by the ability to navigate within the model. For this purpose, CAFM software uses proprietary or externally available BIM viewers as plugins. Depending on the depth of implementation, this can be considered as integration of a BIM model into the software.

Few CAFM systems are able to combine the CAFM data set and the 3D building model in such a way that FM processes can be visualized and controlled via mobile devices such as smart glasses or head-mounted displays (see Sect. 2.6). This would enable a CAFM software, for example, to navigate a maintenance team through the building, provide them with the information they need about technical equipment and the task to be performed at the job site, and process the required feedback directly during task execution.

Solutions of this type currently require intensive cooperation between manufacturers of different hardware and software, which needs a significant development effort. Nevertheless, they outline a viable way, the benefits of which cannot be denied for certain object and operating scenarios.

8.2 CAFM Certification and BIM

In 2008, the GEFMA CAFM Working Group—today GEFMA Digitalization Working Group decided to support GEFMA's quality initiative by contribute to CAFM quality assurance through software certification. In January 2010, the corresponding GEFMA guideline 444 was published for the first time. In addition to the description of the certification process, it contained a collection of 9 catalogs with criteria whose fulfillment should be demonstrated in a software test.

In the following years, both the great need of the users for such a certification and a high acceptance of the procedure by the vendors became obvious. While within two years around 20 vendors had their CAFM products certified, the certification of the software according to GEFMA guideline 444 is now being requested more and more frequently in CAFM tenders. Today, the seal of approval (Fig. 8.3) is regularly part of procurement procedures.

The procedure of certification has now been established for over a decade and is constantly being further developed. Among the currently 17 catalogs of criteria, there has been a catalog since 2018 that also includes Building Information Modeling (NN 2020a). It asks for basic functions in the exchange and processing of BIM data (cf. Fig. 8.4), such as:

Fig. 8.3 Seal of approval for successful certification according to GEFMA guideline 444

Software name:		**Version:**	
Auditors:		**Assessment date:**	
		Duration of assessment:	**ca. 45 min**
Remark	The catalogue does not request BIM functionalities within the CAFM software. Rather, it must be demostrated that imported and exported data can be displayed either in the CAFM database or in the BIM software (authoring tool) before and after transfer. Files to be imported and created during export (IFC or other model-based formats) must be opened with a suitable, freely accessible viewer published by a neutral party and it must be possible to display their contents in a structured manner. GEFMA currently recommends the FZKViewer of the Karlsruhe Institute of Technology (https://www.iai.kit.edu/1648.php).		
	Criteria/Functions		
	Function	**Examples/Data**	
☐	[1] Can a spatial structure together with factual data from the BIM model be transferred to the database of the CAFM software and displayed there?	Buildings, floors, rooms with room numbers and room type	
☐	[2] Can space data from the BIM model be transferred to the CAFM software?	Room area with DIN 277 type of use	
☐	[3] Can graphical/geometric data from the BIM model be visualized in the CAFM software? (with the graphic functionalities from the "Space Management" catalog)	Floor plan	
☐	[4] Can equipment / inventory with essential attributes from the BIM model be transferred to the CAFM software?	Technical equipment, inventory, furniture with manufacturer, model/type no., performance data, etc.	
☐	[5] Can changes in the CAFM system (object data, optionally geometric data) be transferred back to the authoring tool?	Equipment is moved to another room	
☐	[6] Can an IFC file be imported into the CAFM system database and exported from there after processing?		
☐	[7] Is there a function for checking and displaying the files to be exported/imported?	Visual check in the external IFC viewer	

Fig. 8.4 BIM catalog A15 from the GEFMA guideline 444

- Import of geometric data,
- Import/export of space data,
- IFC import/export as well as
- Display and checking of the BIM model.

To check these minimum requirements, the vendor is asked to first open a BIM model (or parts thereof) in an external BIM Viewer and present to the auditors relevant BIM demo objects that need to be modified later on. The model is then imported into the database of the CAFM software. The imported data is displayed and the demo objects are modified. After exporting the modified BIM model, the changes must be visible in the external viewer again.

Fig. 8.5 shows that more than 70% of the currently certified CAFM products have passed the BIM catalog A15 test successfully.

As with the other catalogs, the BIM catalog will be further specified and adapted to current needs as part of the further development of the guideline, taking into account suggestions and requirements from vendors, consultants and users.

No.	Company	Software	Validity	A1 Basic catalog	A2 Space management	A3 Maintenance management	A4 Inventory management	A5 Cleaning management	A6 Space and asset reservation	A7 Lock management	A8 Move management	A9 Rental management	A10 Energy controlling	A11 Safety and occupational health	A12 Environmental protection management	A13 Help and service desk	A14 Budget management and cost tracking	A15 BIM data processing	A16 Contract management	A17 Workplace management
1	ARCHIBUS Solution Centers Germany GmbH	ARCHIBUS V.2022.2	01/2023 - 12/2024	✓	✓	✓	✓	✓	✓	✓	✓	✓	✓	✓	✓	✓		✓	✓	-
2	AT+C EDV GmbH	Facility-Manager VM-7 Version 22	03/2022 - 02/2024	✓	✓	✓	-	✓	-	✓	✓	-	✓	-	✓	-	-	-	-	-
3	Axians Infoma GmbH	Infoma newsystem Liegenschafts- und Gebäudemanagement Version 21.2	01/2022 - 12/2023	✓	✓	✓	-	✓	✓	✓	-	✓	✓	-	-	✓	✓	✓	✓	-
4	Byron Informatik AG	Byron/BIS Version 5.2	07/2021 - 06/2023	✓	✓	✓	✓	✓	✓	✓	✓	-	-	-	-	✓	-	✓	✓	-
5	Dalux Germany GmbH	DaluxFM Version 2022.1	06/2022 - 05/2024	✓	✓	✓	-	-	-	-	-	-	-	-	-	✓	✓	-	-	-
6	EBCsoft GmbH	Vitricon Version 6	04/2021 - 03/2023	✓	✓	✓	✓	✓	✓	✓	✓	✓	✓	✓	✓	✓	✓	-	✓	-
7	Facility Consultants GmbH	getFM Version 6.016	10/2021 - 09/2023	✓	✓	✓	✓	-	-	✓	-	-	-	-	-	✓	-	-	✓	-
8	Hexagon	HxGN EAM Version 11.7	08/2022 - 07/2024	✓	✓	✓	✓	-	✓	-	-	-	✓	✓	✓	✓	✓	✓	-	-
9	HSD Händschke Software & Datentechnik GmbH	HSD NOVA FM Version 4.0	03/2022 - 02/2024	✓	✓	✓	✓	✓	✓	✓	-	✓	✓	✓	✓	✓	✓	✓	✓	-
10	IBM GmbH	IBM TRIRIGA, Plattform 4	12/2022 - 11/2024	✓	✓	✓	✓	✓	✓	✓	✓	✓	-	✓	✓	✓	✓	✓	✓	✓
11	Iffm GmbH	iffmGIS Version 15	10/2021 - 09/2023	✓	✓	✓	✓	✓	-	✓	-	✓	✓	-	-	✓	✓	-	✓	-
12	InCaTec Solution GmbH	Axxerion, Version 22	04/2022 - 03/2024	✓	✓	✓	✓	✓	✓	✓	-	✓	-	✓	✓	✓	✓	-	✓	-
13	Keßler Real Estate Solutions GmbH	FAMOS 4.6	03/2023 - 02/2025	✓	✓	✓	✓	✓	✓	✓	✓	-	-	✓	✓	✓	✓	✓	✓	-
14	KeyLogic GmbH	KeyLogic, Version 2022 Q3	08/2022 - 09/2024	✓	✓	✓	-	-	-	-	-	-	-	-	-	-	-	-	-	-
15	KMS Computer GmbH	GEBman-10 Version 9.1	07/2021 - 06/2023	✓	✓	✓	✓	✓	✓	✓	✓	✓	✓	✓	-	✓	✓	✓	✓	-
16	Loy & Hutz Solutions GmbH	waveware V11.200	06/2022 - 05/2024	✓	✓	✓	✓	✓	✓	✓	✓	✓	✓	✓	✓	✓	✓	✓	✓	✓
17	facility (24) GmbH & Co. KG	facility (24) - Version 5	01/2023 - 12/2024	✓	✓	✓	✓	✓	✓	✓	✓	✓	✓	✓	✓	✓	✓	✓	✓	✓
18	N+P Informationssysteme GmbH	SPARTACUS Facility Management 4.5.4	12/2022 - 11/2024	✓	✓	✓	✓	✓	✓	✓	✓	✓	✓	✓	-	✓	✓	✓	✓	-
19	pit - cup GmbH	pit - FM 2021	06/2023 - 05/2025	✓	✓	✓	✓	✓	✓	✓	✓	✓	✓	✓	✓	✓	✓	✓	✓	-
20	Planon GmbH	Planon Live / Version L82	07/2022 - 06/2024	✓	✓	✓	✓	✓	✓	✓	✓	✓	✓	✓	✓	✓	✓	✓	✓	✓
21	RIB IMS GmbH	iTWOfm V13	01/2023 - 12/2024	✓	✓	✓	✓	✓	✓	✓	✓	✓	✓	✓	✓	✓	✓	✓	✓	✓
22	sMOTIVE GmbH	OSIRIS 2023	04/2023 - 03/2025	✓	✓	✓	✓	✓	✓	✓	✓	✓	✓	✓	✓	✓	✓	-	✓	✓
23	VertiGIS GmbH	ProOffice 9.1	07/2021 - 06/2023	✓	✓	✓	✓	✓	✓	✓	✓	✓	✓	✓	-	✓	✓	✓	✓	-

Fig. 8.5 Overview of certified CAFM products according to GEFMA 444 (as of 01.06.2023)

8.3 BIM and ERP Systems

Operation of a large company without an Enterprise Resource Planning (ERP) system is almost impossible today. The complexity and diversity of business processes can only be mastered with structuring and agreeing on common standards. Collaboration between departments and business partners must take place smoothly, which can only be achieved through uniform formats for exchanging information and a "single source of truth" (often integrated into the ERP system).

This can be transferred to the BIM approach. Here, too, it is necessary to simplify processes through common standards and the common access to the corresponding data models. The interlocking of BIM and ERP, possibly within a CDE, promises a beneficial combination of tasks during planning, construction and operation of real estate.

At first glance, BIM models have few or no similarities with the numbers-based economic perspective of ERP systems. BIM models describe real estate as realistically as possible with regard to their construction and technical view, ERP systems always support the business processes around real estate in relation to a property and/or a building/component or an associated technical asset. On a second look, it therefore makes sense to identify the points of contact between the two worlds and to describe the resulting commercial follow-up processes. Therefore, it is necessary to discuss the connection of both perspectives with BIM tools (such as authoring tools) as a data source in a more structured way.

An example is given below, in which the use of the BIM approach promises significant benefits.

The cooperation between owners and facility managers is usually characterized by a regular exchange of building-related data. This exchange often takes place across corporate and system boundaries. Even though this has become much easier in recent years due to different approaches and the introduction of cloud solutions, data consistency is still a major challenge.

The future use of a single BIM model with a standardized description and structure will significantly improve the data quality and exchange of information. For example, if all parties involved are given access to the digital twin (see Sect. 4.1), this will become a permanent part of the work of owners as well as facility and property managers.

8.3.1 ERP

An ERP system typically covers important business processes of a company. These include in particular the areas:

- Finance and accounting,
- Controlling,
- Materials management,
- Production,
- Human resources management,
- Research and development,
- Sales and marketing,
- Maintenance and repair.

The integration of these areas and the avoidance of isolated solutions in one IT system are often referred to as a holistic ERP system or an integrated ERP system.

An ERP system is the digital representation of the company (organization and processes)—just like BIM provides a digital representation of real estates. During the operation of buildings, most of the disciplines of an ERP system are used to optimally control properties. A linking of the commercial processes with a BIM model therefore appears to be obvious.

The use of the BIM approach in the operation of buildings has an impact on commercial and technical processes within the ERP system. Differences in ERP systems can be found in the coverage of property-specific processes, such as contract management, maintenance and utility billing. Some providers do not offer any or only rudimentary functions for this, others offer integrated functions. Consequently, when an interaction between BIM and ERP systems is considered, the range of functions of the ERP solution must be considered as well.

8.3.2 Use Cases

When considering the benefits of the BIM approach in operations of real estate, the identification of value-added applications (Use Cases) is key. The application areas are very diverse, as most real estate-related processes are based on building data. In technical building management, the relationship between consistent and up-to-date data for maintenance, modernization and operational safety is obvious. Also in the subsequent commercial processes such as purchasing services as well as letting or renting, BIM data reduce the costs significantly.

First, the relevant processes or use cases are identified and described, in which a BIM-ERP cooperation or integration is beneficial. In the second step, the IT-technical implementation is explained. Use cases arise from the processes of building and facility management, as they are described, for example, in DIN 32736 "Building Management—Terms and Services" (NN 2000) and in the FM processes according to GEFMA guideline 100-2 (NN 2004).

8.3.2.1 Maintenance and Repair

In the context of maintenance and operations, links are required between the master data objects and their attributes on the one hand, and the measures on the other hand. The master data objects contain information on the frequency, type and implementation of measures. This information is a basis for maintenance planning and the execution of the measures.

In the BIM perspective, maintenance objects are no longer represented only as alphanumeric objects with attributes and possibly attached 2D drawings, but as 3D objects that are located within the digital twin (cf. Sect. 4.1), for example in a room. The navigation to the data record of the object to be maintained is no longer via object hierarchies, but by means of a digital "walk through" or based on a freetext search (Google-like search). Maintenance and operations schedules can already be linked to the objects in the planning phase and transferred to the BIM model when transitioning from the construction to the operational phase. Maintenance schedules can be loaded and updated directly from cloud-based platforms (e.g. Honeywell Forge platform, NN 2021m). Based on this data, measures for maintenance and operations are planned in the ERP system or created ad hoc in the event of a fault. The classical functions of the ERP system are used for this purpose. In the case of outsourcing services, materials, personnel and, if necessary, machine and asset costs are planned and ordered via the procurement processes represented in the ERP system (invitation to tender and individual contract or framework contract) and finally procured. After execution, the internal or external FM service provider reports back the consumed materials and provided services in order to initiate checking, release and settlement of the invoice.

In the event of a warranty claim, the link between maintenance objects and the BIM model can be used to read out the warranty conditions and deadlines from the BIM model. Hence, the specification of warranty periods by the executing companies is a prerequisite for new construction or renovation, refurbishment and repair.

8.3.2.2 Modernization and Conversion

The processes of modernization and conversion have their origin in the planning of respective measures. The planning is implemented based on data from the digital twin in the BIM model. Based on the planning, bills of material and services are created, e.g. by connecting a system for tendering, awarding and billing. The bill of services created in this way form the basis for tendering and awarding of the services. With the award of the contract to the executing companies, the integration into the ERP system is completed, because the commissioning or ordering has to be mapped in the ERP system from a commercial point of view (creditor, delivery conditions and place for delivery, payment conditions, service items, discounts, taxes, accounting objects).

In order to cost control the measures during the construction phase, a corresponding object structure has to be mapped in the ERP system. In Germany, this is either based on DIN 276 (NN 2018b) or on the trade structure of the standard services book construction of GAEB (NN 2021q). The costs are planned in this structure during the planning phase of the measure. The planning is based on the bill of services mentioned above. In the following phases up to the execution, the planned costs are updated. In the execution phase, the last planning version serves as the basis for the target-actual comparisons and cost control of the execution.

During the execution of the services, the planning is also updated as an as-built documentation in the digital twin of the BIM model. Costs arising from interim invoices of the executing companies are recorded, checked and released for payment in the ERP system.

Further details on the processes of construction management from the perspective of the client and the necessary IT support can be found in Limberger (2005).

8.3.2.3 Space Management

According to DIN 32736 (NN 2000), space management comprises, in addition to space and area analysis, vacant space management and other space-related services such as catering, cleaning and security, in particular occupancy control. These include, among other things, the following areas of responsibility:

- Permanent assignment as a fixed-desk or flex-desk assignment,
- Provision of short-term assignable spaces, e.g. for meeting and conference rooms,
- Variant planning for moves.

Space management is based on floor plans of the respective buildings, with techniques such as stacking and blocking also being used. As shown in Fig. 8.6, the floor plans can be generated as 2D extracts of the BIM model for the individual floors of the build-

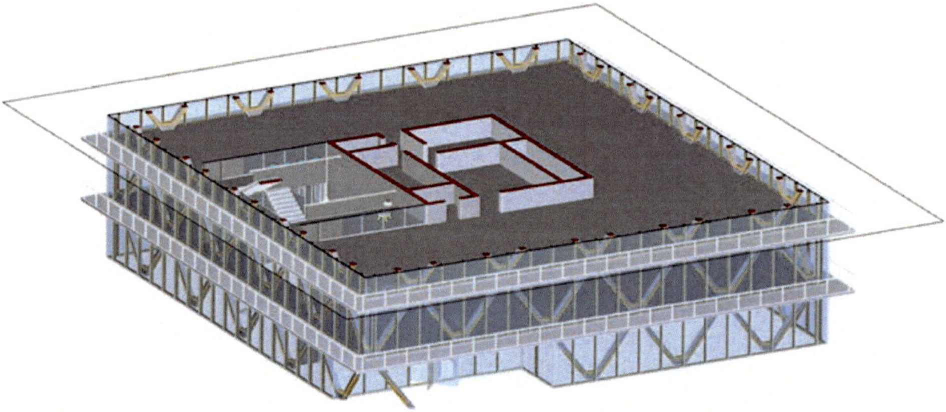

Fig. 8.6 Cross section through a digital 3D model for generating a floor plan for occupancy planning

ing. Based on the floor plans, the assignable spaces and rooms are then determined and assigned to the users. Depending on the company, there are different working methods, which in turn require different levels of integration into the ERP world.

1. *Assignment of floor spaces to individual organizational units*
   In this approach, areas are marked in the floor plan and are assigned to an organizational unit, e.g. cost center or project. A more detailed planning of the occupancy of the individual workplaces is not done. Detailed planning is the responsibility of the organizational unit that is supposed to use the workplaces. This approach is typically used when mostly adjacent flex-desk- or shared-desk zones (home zones) are to be mapped, because a detailed assignment of the employees to individual workplaces is only possible to a limited extent, as the assignment can change on a daily base. In this approach, organizational data such as the number and description of cost centers using the ERP system are required to assign it to the space.
2. *Assignment of spaces to individual occupants*
   In this approach, the floor plan is supplemented with representations (symbolic or architectural) of the individual workplaces. These workplaces are assigned to employee master data using either clear or anonymous names. Multiple occupancy scenarios can also be mapped. In this approach, an integration into the human resources module of the ERP system is necessary. In this way the employee master data can be extracted for the assignment in the occupancy plan and the room assignment can be transferred back into the personnel master data. A link with external human resources management systems is also possible if these are already in use. For the processes of cost allocation described in Sect. 8.3.2.4, it is also possible to map the organizational assignment of employees in the occupancy data, if this is permitted in the respective company. Space management for meeting and conference rooms is

often controlled via reservation systems, in which users can book rooms. For booking, either the employee number, email address or the number of the organizational unit to which they belong is necessary. The rooms are usually occupied for hours or days, but the current and future occupancy situation is not displayed in the floor plan. In the floor plans, the rooms are only marked as temporarily occupiable so that they cannot be used by permanent occupancy scenarios. Further requirements of meeting room management, such as booking services for catering, technical equipment or seating, are not considered here (see also Sect. 10.2).

The sub-processes of move management range from variant planning of future occupancy to planning of logistic processes for the move, e.g. how many boxes and pieces of furniture have to be moved from A to B. Here, only the sub-processes of variant planning are considered within the scope of occupancy planning. Move management is based on the information of the occupancy management. The task of move management is planning and evaluation of variants for future occupancy depending on the specific planning described above. The floor plans obtained from BIM model are used to visualize the occupancy scenarios for the comparison of variants.

8.3.2.4 Commercial Real Estate Management

With regard to the integration of BIM models and ERP systems, the following processes of commercial real estate management play a minor role because they do not directly access and further process data from the BIM model, as is the case, for example, with floor plans in occupancy management. These commercial real estate processes either support the activities mentioned in the previous sections in the sense of end-to-end software support or, as in the case of leasing, where data from the BIM model is used to describe the leased items in more detail. The processes of commercial real estate management that have no relation to BIM data, such as price adjustment or turnover rent calculation, are not described here.

1. *Leasing*

   During renting and leasing, data from BIM models is used to create floor plans. The floor plans are used for the following tasks:
   - Determination of rental and leasing spaces,
   - Representation of the room layout and the supporting components,
   - Planning of tenant installations,
   - Creation of evacuation plans.

   The floor plans are usually provided as attachments to the rental or lease agreement, either in CAD or PDF format, and stored in the electronic file of the contract. In the case of leasing, the lessees use the plans provided by the lessor for further processes, such as space management. However, these plans are decoupled from the original

BIM model and are therefore not subject to the continuous update service, as is the case for BIM models.

2. *Asset accounting*
   If links to BIM models are used in asset accounting, they serve to visualize the assets, e.g. the building or the technical equipment installed in the building. So far, data is not or only rarely downloaded from the BIM model to set up the master data in asset accounting, as the master data required in asset accounting is only partially stored in the BIM model.
3. *Cost allocation* and *Service charge calculation*
   Cost allocations are often used in corporate real estate scenarios. Here, real estate and FM provide spaces for other operational and administrative units of the company. FM charges the costs for maintenance and operation of the real estate to the using units. The charge is either made on the basis of actual costs or standard costs. The charge and distribution key is either the space used by the respective organizational unit (ratio of the space used to the total usable space of the building, applied to the total costs) or the number of occupants in relation to the total number of building users. In rare cases, meters or sensors are used if they are available and legally permissible.
   In the case of charges based on occupied or usable space, these are determined from the BIM model and transferred to the ERP system either on the basis of individual rooms or as a total sum (cf. Sect. 8.3.2.3).
   This data use from BIM models also applies to the use of charge-relevant space data required for service charge statement. Further sub-processes of service charge settlement, such as the determination of the amount of advance payments or flat-rate payments or the calculation of credits and claims, do not offer any potential for BIM integration.
4. *Purchase* and *Procurement*
   Purchasing and procuring are support processes for the core processes of construction, renovation, building maintenance, and repair.
   The needs covered by purchasing and procurement always concern components and object groups of the BIM model.
   In the case of new construction, renovation and refurbishment measures, component and service lists are generated from the BIM model and passed on to the procurement process. There the sub-processes of preparation and implementation of the award, the commissioning and finally the invoice verification are carried out during execution.
   Maintenance and repair measures are usually smaller activities in terms of duration, type and volume compared to new construction and renovation. Maintenance is planned in ERP systems regarding priority, volume (demand for materials and personnel) and are automatically triggered, i.e. either own personnel is scheduled and commissioned or a service provider is commissioned. The commissioning of external service providers is usually executed based on framework agreements (often service

level agreements), in which the services, billing rates, materials and costs as well as response times are specified. The BIM model provides the technical asset data including components and assemblies as well as the required parts and materials.

8.3.3 IT-Technical Implementation

The technical implementation of IT is depicted in Fig. 8.7 (Limberger 2020). It shows the necessary components of an ERP software for real estate companies with the BIM model as digital twin at its core. Since ERP systems usually do not have a standard BIM component and on the other hand there are many software solutions with which BIM models can be generated and edited, the central question of the IT-technical implementation is the integration of the different solutions via interfaces. It also has to be clarified whether a real data exchange or referencing makes more sense for the process to be supported.

Without processes being in place to manage the updating of information and models organisations always bear the risk of data aging. However, it offers the advantage that extracts from the BIM model, such as floor plans, can be used without the BIM model being available (e.g. the BIM server is offline).

Referencing has the advantage that no local copies have to be maintained and updated. However, processing speed can be reduced for complex operations because reading from a central BIM server is slower than reading from a local copy.

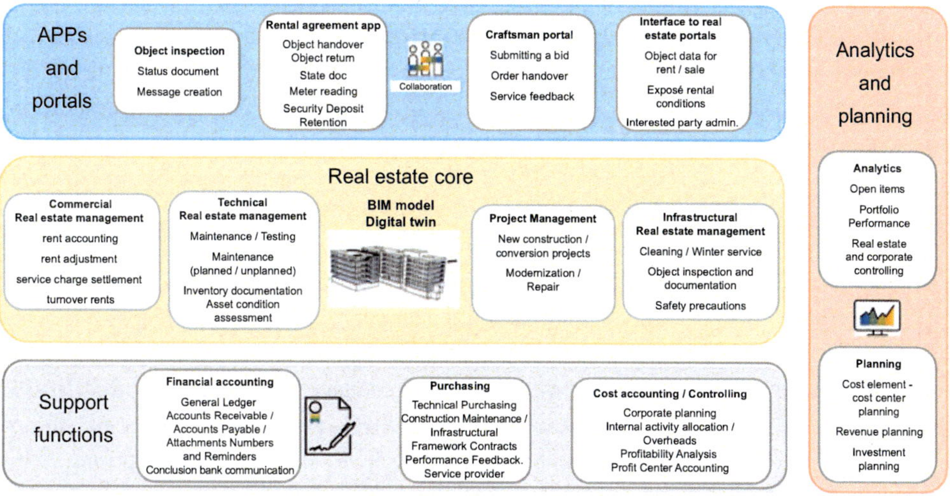

Fig. 8.7 Structure of an ERP solution for real estate companies

The basis for open data exchange is ISO 16739-1 (NN 2018a). This standard regulates uniform exchange of BIM models between different software systems.

The description of the integration scenarios is based on the use cases described in Sect. 8.3.2.

8.3.3.1 Implementation: Maintenance and Repair

The technical implementation of this use case with IT relates to both the master data maintenance (initial creation and updating) in the ERP system and the support of processes such as creating a maintenance work order.

A copy of the alphanumeric properties and variants including the structural hierarchy and ID is necessary for master data maintenance. This allows a relationship to be established between the component in the BIM model and the data record in the ERP system. Mutual data enrichment can then take place via this assignment, such as:

- Display of open and closed fault messages and repair orders in the BIM model by mouse-over or click-on links,
- Visual 3D representation or exploded view of components as an supplement of the alphanumeric ERP data,
- Use of the BIM model supporting indoor navigation for technical personnel to find the installation site in the building more easily,
- Linking to additional component attributes, such as technical specifications or work instructions, as a work aid for technicians and craftsmen.

In addition, component descriptions can be used to determine consumables and spare parts that are needed in maintenance and repair measures and to transfer them to the maintenance order afterwards. Standardized data formats for this purpose are described in the next section.

8.3.3.2 Implementation: Modernization and Renovation

This use case not only concerns the execution planning of measures and the associated change of the BIM model, but above all creation of service specifications, awarding of the contract and monitoring of the execution. The most important element is the service specification generated based on the component information from the BIM model. Schiller and Faschingbauer (2016) describe the data exchange, which is based on DIN SPEC 91400 (see NN 2017c). It should be noted that there are a large number of national, European and international standards for component and trade descriptions, which need to be harmonized. According to Schiller and Faschingbauer (2016), this has been achieved with the establishment of the IFC standard and DIN SPEC 91400, although in practice proprietary data exchange formats such as Revit's rvt are also used.

In Germany, it can be generated on the basis of the GAEB standard (NN 2021p) and exchanged with execution companies without media disruption. With the GAEB standard for time-based work (STLB-BAUZ) (see NN 2021r), this is also possible for framework contracts, as used for maintenance and repair work.

The service specification is then used in the ERP system to process invitation to tender, award of contract and billing as well as payment, that are controlled by specific software systems. Those systems can either be fully or partially integrated into the ERP solution. Payment of invoices is always triggered in the ERP system.

It is necessary that the BIM model has been updated accordingly with the planning data before the service specifications are created.

More complex modernization and renovation measures are often processed in ERP systems in their project management components or modules. For this purpose, project structure plans are created for the individual trades and services. These are in turn linked to the objects of the purchasing module (order/commission) to enable continuous cost planning, budgeting, obligo updating and actual cost tracking. The individual elements of the project structure plan can be linked to the components of the BIM modelgiving the project management a better picture of the task.

8.3.3.3 Implementation: Space management

The basis for the use cases of space management are up-to-date space information. These are also extracted from the IFC BIM model and are initially transferred to the ERP system in alphanumeric form (room stamp information) and updated in the event of a change. Again, a local copy of the BIM component is created in the ERP system. Likewise, the floor plan generated from a horizontal section is stored in the ERP system as a copy.

A copy of the room master data in the ERP system is required for occupancy and billing processes as well as for alphanumeric floor space balance and KPI's such as occupancy rate, vacancy rate, cost per space unit. Room data in connection with occupancy data represent mass data, which can be processed more efficiently if a copy is created in the ERP system.

The floor plans are needed for graphical reports of the occupancy and utilization situ ation as well as for the variant planning of moves. The floor plan with the corresponding CAD layers is enriched with additional layers that display ERP data.

It should be mentioned that, especially in the corporate real estate sector, landlords often do not provide a BIM model for rented space, but only simplified 2D plans, which often are not vector graphics.

8.3.3.4 Implementation: Commercial Real Estate Management

The use cases of commercial real estate management for renting and leasing as well as service charge billing rely on space and room data. How these are extracted from the BIM model and maintained as a copy in the ERP system has already been described in the previous section.

The same applies to the processes for purchasing and procurement. In Sect. 8.3.3.2 it was described how a bill of quantities can be generated from BIM component information. Creating a bill of quantities from the perspective of purchasing is just a preliminary activity.

The BIM integration into asset accounting is a rare use case, which can be used to visualize and/or locate assets in the building. Here, references or linkages are used.

8.4 Cooperative Platform Concepts as CDE

The previous chapters have described important instruments of the BIM method and presented solutions for their use in FM. However, it is known from practice that many companies still complain about a deficient data basis for their building-related information.

Even if a good basis has been laid by BIM in the planning and construction phases, the lack of standardization of data and information flows in the operational phase still leads to the formation of data silos in many companies. As a result, the implementation of seamless processes, informed decisions and "smart" data applications is often difficult or even impossible. This is especially true if operators of large portfolios also have to deal with internationally divergent requirements and conditions.

8.4.1 A Data Model for the Real Estate Industry

A uniform, open data standard that can be used by everyone not only simplifies the implementation of an asset information system described in Sect. 4.3. It also forms an essential basis for the use of advanced technologies such as machine learning and artificial intelligence. The data exchange formats established so far by BIM reach similar limits as the BIM methodology itself. Currently, they focus content-wise on the project phases of planning and construction. For data that arises during the operational phase in the management of buildings, there have only been a few established comprehensive approaches with an international focus.

The International Building Performance & Data Initiative (IBPDI), founded in 2020, aims at developing an international uniform data language and semantics for the real estate industry based on standards already available and used on the market. IBPDI (cf. NN 2021t) has summarized complex building data and data for technical and commercial building operation, including energy and CO_2 consumption data, in a comprehensive data model (Common Data Model). This is divided into various clusters (cf. Fig. 8.8), with the clusters particularly relevant for BIM and FM shown in orange color.

The *Digital Building Twin* cluster is based, among other things, on the international COBie standard (Construction Operations Building Information Exchange). Other standards integrated are IFC, CAFM-Connect, ISO 81346, DIN 276, ebkp-h and IPMS.

The *Energy & Resources* cluster, based on CRREM, GHG Protocol Corporate Standard and GRESB, also includes aspects of the application of circular economy criteria, including renovation, conversion and retrofitting activities.

The *Property Management* cluster was extended by a large number of data entities based on gif and RICS standards—for example, on the topic of tenant relationships. In addition, the cluster *Portfolio and Asset Management* based on gif and RICS standards was linked to the *Financials* cluster, allowing reporting based on uniform key figures (Key Performance Indicators—KPIs).

The *Financials Cluster* is based on international accounting standards such as the International Financial Reporting Standards (IFRS) or the MSCI World Index, aiming to represent the development of global stock markets and the world economy in general. The cluster includes a generic chart of accounts that allows for the analysis of standardized financial metrics across industries. Another focus is on lifecycle costs, which are defined in detail by RICS.

The *Common Data Model (CDM)* for Real Estate is continuously developed by IBPDI members, including numerous representatives from the real estate industry as well as associations such as GEFMA. The current focus is on the *Facility Management* and *Organizational Management* clusters, in which the entities of the people and organizational perspectives in the building context and, in particular, the processes and roles of the value chain are defined.

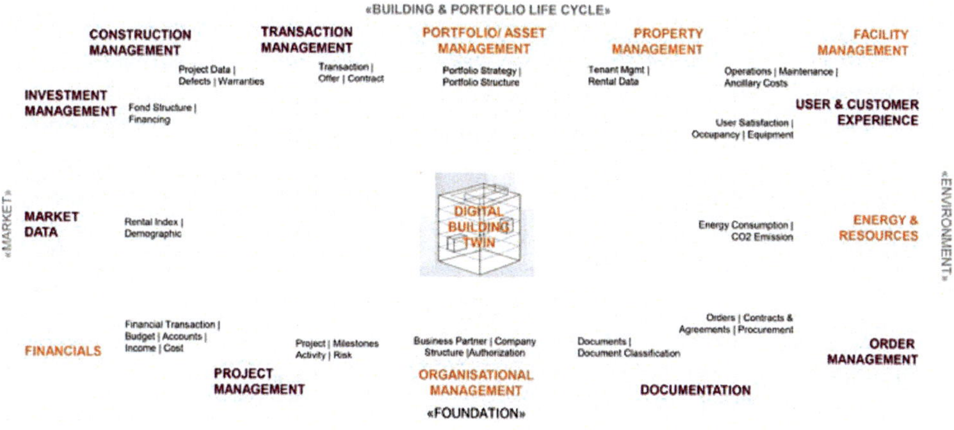

Fig. 8.8 Overview of the IBPDI data clusters

The CDM for Real Estate is an open data model accessible to all players in the real estate industry that can be used as a basis for storing and exchanging data within the entire real estate sector. The idea of networking concept and integrative collaboration shaped by the BIM methodology could thus be implemented in the future for all relevant real estate and FM processes (cf. NN 2021t).

8.4.2 The Use of Platforms in FM

In addition to the technical solutions of an asset information model (AIM) as a Common Data Environment (CDE) for the use phase, presented in Sect. 4.3, the organizational question remains as to which party is responsible for managing the data based on the CDM discussed in the previous section, once the developer or general contractor has handed over CDE responsibility after completion of the construction phase.

In the context of CAFM systems, a problem often arises in practice which also highlights a limit for cross-stakeholder CAFM-BIM integration—the contractual obligation of the FM service provider to update building data during operation. Depending on the contractual terms, this obligation to data update can either take place in the client's system or in the FM service provider's system.

Both variants have advantages and disadvantages, which also affect the maintenance of BIM data. If data updating (including BIM models) takes place in the client's systems, this leads to duplicate work for the service provider, who must maintain at least some data in his own systems for controlling his own employees. Those systems are, for example, required for billing and working time recording. However, if data updating only takes place in the systems of the FM service provider, a building owner has the risk of a time-consuming data migration or even data loss when changing the service provider. A possible solution to this dilemma is data exchange via an intermediate platform, which offers all stakeholders around real estate management simultaneous data access. Technically, such a platform largely corresponds to the explanations in Sect. 4.3.2.4, with a CDE from the planning and construction phase now also involving the following stakeholders (see Fig. 8.9):

- Corporate Real Estate Management,
- Asset Management,
- Property Management,
- Facility Management,
- Purchasing,
- Accounting and financial administration,
- End users (employees of the client, tenants, visitors/patients),
- Service providers,
- Consultants and service controllers.

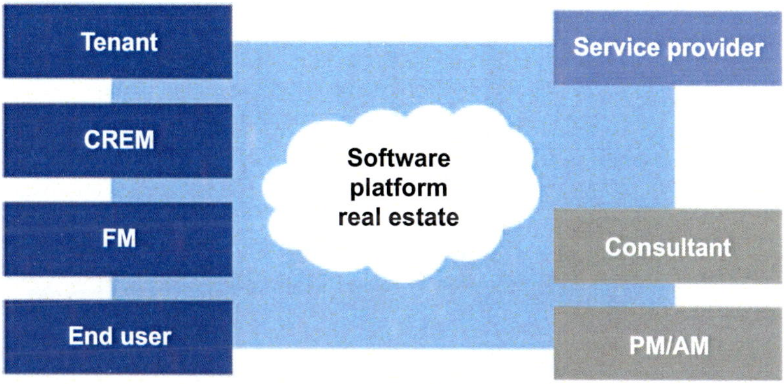

Fig. 8.9 Key stakeholders in real estate management

In addition to the providers of CDE's for the construction phase, more and more platform solutions are emerging for real estate operations. Unlike existing applications, which usually focus on one party (building owner/operator or service provider), these platforms enable, depending on their architecture, collaboration between two or more parties involved in a property. This means that data can not only be collected and managed, but also distributed or even enriched by different parties.

Other applications can be connected to such platforms via open interfaces (APIs) and, from the perspective of the building owner/operator, also external service providers can be integrated into the collaboration around the building (cf. Fig. 8.10). These include, for example, building management systems and, increasingly, IoT sensors, the focus of which is the measurement of the building condition (e.g. temperature, light, noise, occupancy and CO_2). The computing power required for such a collaborative platform can be covered scalably by cloud technologies available today. This enables to actually process data from technical assets and relate it to each other.

This helps to optimize FM using real-time information and to gain direct insight into building operation, such as cost control, user satisfaction and workplace management. Monitoring of technical building equipment and assets, space optimization and space utilization, as well as improvement of comfort factors are typical applications. Other functions such as adding and linking documents to the building or building parts based on metadata, as well as activity management, make the platform a reliable source of information (single source of truth) for all stakeholders.

8.4.3 Status Quo and Outlook

Such a technological platform offers building owners and operators of large portfolios the opportunity to interlink information from previously independent and heterogeneous system environments and at the same time to involve their external stakeholders (e.g. service providers, potential buyers, tenants).

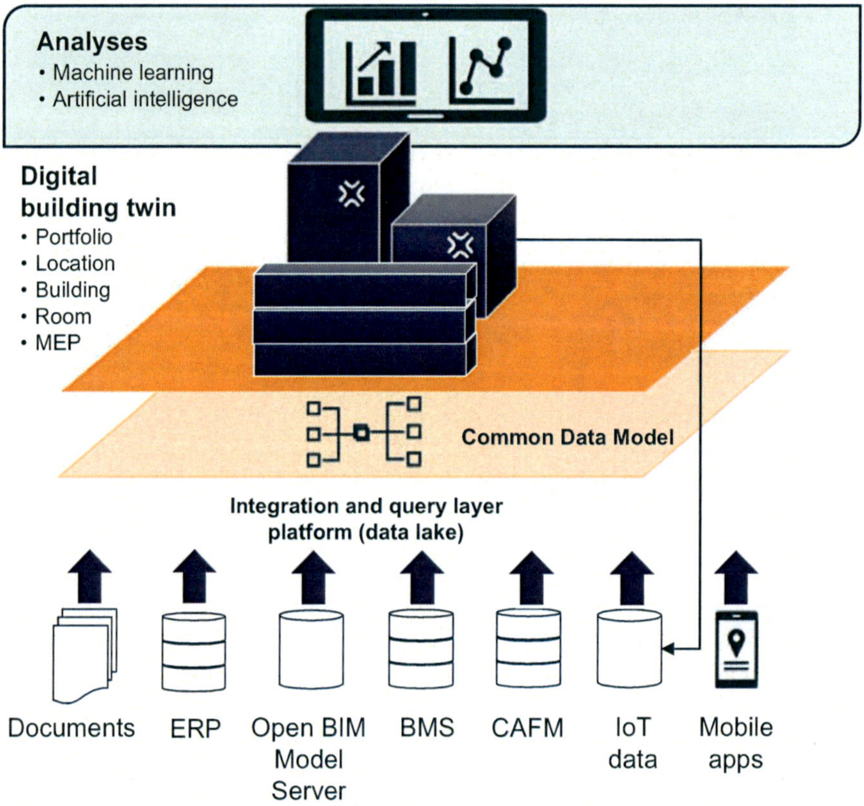

Fig. 8.10 Integrated asset information system based on cloud platform technology using a unified data model

The acceptance of utilizing platforms has arrived on the market, but market participants still see a number of hurdles.

One of these has been the lack of existing data standards (Ball 2018). Initiatives such as the aforementioned IBPDI can help overcome that.

Other hurdles include data protection, IT governance, complexity of the service description and lack of trust. In addition, the coordination with the IT responsible staff in the companies regarding the use of digital platforms has not yet been carried out (Berger 2020).

Hence, there is still no final picture of the future strategic role of platforms on the market. However, the integration of data creates a great potential for new business models. Accordingly, the range of services and functions is growing rapidly. It is to be expected that step-by-step establishment of platforms in the operational phase will also lead to a change in the digital support of real estate and FM. which will have a positive impact on disseminating the BIM method in the operational and use phase.

8.5 Summary

Increasingly, real estate and facility managers have recognized the value that data from BIM models during the operation of buildings can offer. In response, CAFM software providers are increasingly providing interfaces to BIM software that enable exchanging data with these systems. This development has also been promoted to a significant extent by national and international associations. One initiative is testing BIM data exchange in CAFM software in accordance with the German GEFMA guideline 444 since 2018. To date, about 70% of all certified systems have been able to successfully demonstrate the functions from the corresponding BIM test catalog. This indicates that BIM has arrived in the RE and FM sector. In this context, CAFM systems support both Open-BIM and Closed-BIM formats. However, some systems also interact directly with BIM authoring tools.

Since ERP systems are in use in almost every large company or public institution, the question of advantages of collaboration between ERP and BIM is increasingly being asked. ERP systems preferably map the organization and business processes, while BIM provides a digital representation of the buildings. Benefits arise in particular where access from the ERP system to BIM data or to a digital twin is required or useful. For this reason, use cases have been described in detail in the previous sections, in which data exchange between BIM and ERP leads to more efficient processes and value added. In addition, it was described how these use cases can be implemented in a BIM/ERP environment.

Finally, the advantage of using cooperative, often cloud-based platforms (in particular Common Data Environments) was discussed. The importance of standardized models for data exchange in the entire FM and real estate ecosystem is pointed out. The Common Data Model from IBPDI is presented as an example. Via application programming interfaces (API's), it is also possible to connect other software applications to such platforms. These platforms will in future offer the integration of data from heterogeneous systems and to analyze them comprehensively.

The further success of this development is also dependent on the willingness of owners and users to compensate for the additional effort that ultimately makes the value of a BIM-CAFM/ERP integration possible.

References

Ball T (2018) Lünendonk-360-Grad-Incentive 2018—Digitalisierung in der Immobilienwirtschaft, Lünendonk & Hossenfelder GmbH, 42 p

Berger R (2020) Inhalte und Nutzen einer digitalen Plattform im Corporate Real Estate und Facility Management. Studie Ronald Berger GmbH und RealFM e. V., 15. July 2020

Limberger B (2005) Unterstützung der Baumanagementprozesse von Immobilienunternehmen mit integrierten betrieblichen Informationssystemen—ERP-Systemen. PhD thesis, Lehrstuhl für Bauwirtschaft, Bergische Universität Wuppertal, DVP-Verlag

Limberger B (2020) Kaufmännische Immobilienverwaltung mit ERP Systemen, Lecture notes SRH Hochschule, Heidelberg, chair Prof. Meysenburg

NN (2000) DIN 32736: Gebäudemanagement: Begriffe und Leistungen, 2000–08, 8 p

NN (2004) GEFMA Richtlinie 100-2: Facility Management—Leistungsspektrum, Juli 2004, 36 p

NN (2017c) DIN SPEC 91400: Building Information Modeling (BIM)—Klassifikation nach STLB-Bau. Deutsches Institut für Normung, 2017-02, 16 p

NN (2018a) ISO 16739-1: Industry Foundation Classes (IFC) for data sharing in the construction and facility management industries—Part 1: Data schema. International Organization for Standardization, 2018–11

NN (2018b) DIN 276: Kosten im Bauwesen. Deutsches Institut für Normung, 2018–12

NN (2020a) GEFMA Richtlinie 444: Zertifizierung von CAFM-Softwareprodukten. Februar 2020, 21 p

NN (2021b) Marktübersicht CAFM-Software. GEFMA 940, Sonderausgabe von "Der Facility Manager", FORUM Zeitschriften und Spezialmedien GmbH, Merching, 2021, 198 p

NN (2021m) https://www.honeywell.com/us/en/honeywell-forge/buildings (retrieved: 07.08.2021)

NN (2021p) GAEB Datenaustausch XML. Gemeinsamer Ausschuss für Elektronik im Bauwesen. https://www.gaeb.de/de/produkte/gaeb-datenaustausch/ (retrieved: 20.08.2021)

NN (2021q) Standardleistungsbuch Bau. Gemeinsamer Ausschuss für Elektronik im Bauwesen. https://www.gaeb.de/de/stlb-bau/ (retrieved: 20.08.2021)

NN (2021r) Standardleistungsbuch für Zeitvertragsarbeiten. Gemeinsamer Ausschuss für Elektronik im Bauwesen. https://www.gaeb.de/de/produkte/gaeb-datenaustausch/ (retrieved: 20.08.2021)

NN (2021t) IBPDI—International Building Performance & Data Initiative. https://ibpdi.org/ (retrieved: 21.09.2021)

NN (2022b) Marktübersicht CAFM-Software. GEFMA 940, Special issue "Der Facility Manager", FORUM Zeitschriften und Spezialmedien GmbH, Merching, 2022, 202 p

Schiller G, Faschingbauer G (2016) Die BIM-Anwendung der DIN SPEC 91400. DIN e. V., Beuth Verlag, Berlin—Wien—Zürich, 88 p

BIM in Real Estate and Facility Management—Case Studies

9

Maik Schlundt, Simon Ashworth, Thomas Bender, Asbjörn Gärtner, Michael Härtig, Reiko Hinke, Markus Krämer, Michael May and Matthias Mosig

9.1 Overview of Case Studies

The GEFMA working group Digitalization has collected and processed practical examples since the development of the first BIM White Paper, so that 11 national and international projects can now be presented in the following sections. Very different application areas were chosen, such as:

- Media,
- Pharma,

M. Schlundt (✉)
DKB Service GmbH, Berlin, Germany
e-mail: maik.schlundt@dkb-service.de

S. Ashworth
Zürcher Hochschule für Angewandte Wissenschaften (ZHAW), Wädenswil, Switzerland
e-mail: ashw@zhaw.ch

T. Bender
pit – cup GmbH, Heidelberg, Germany
e-mail: thomas.bender@pit.de

A. Gärtner
IU Internationale Hochschule, Erfurt, Germany
e-mail: a.gaertner@iubh-fernstudium.de

M. Härtig
N+P Informationssysteme GmbH, Meerane, Germany
e-mail: michael.haertig@nupis.de

M. May et al. (eds.), *BIM in Real Estate Operations*,
https://doi.org/10.1007/978-3-658-40830-5_9

- Bank,
- Museum,
- Technology park,
- Energy supplier,
- Airport,
- Administration,
- Municipality.

The purpose of the case studies is to show the practical implementation of BIM with added value for real estate (RE) and facility management (FM). From these examples, experiences and insights as well as suggestions for own BIM/FM projects can be drawn. Realistic requirements for a successful technical implementation can also be derived from these examples.

The case studies, among other things, shed light on the following points:

- Introduction of company and participants,
- Project presentation,
- Objectives of the use of BIM and implementation procedure,
- Software and functionality used,
- Results and findings,
- Added value of the integration of BIM and CAFM.

This should provide a better understanding of BIM in RE and FM and show the possibilities but also the limitations.

R. Hinke
BASF SE, Ludwigshafen, Germany
e-mail: reiko.hinke@basf.com

M. Krämer
Hochschule für Technik und Wirtschaft Berlin, Berlin, Germany
e-mail: markus.kraemer@htw-berlin.de

M. May
Deutscher Verband für Facility Management (GEFMA), Bonn, Germany
e-mail: michael.may@gefma.de

M. Mosig
TÜV SÜD Advimo GmbH, München, Germany
e-mail: matthias.mosig@tuvsud.com

9.2 Municipal Real Estate Jena

9.2.1 The Project

An example of a successful project is the municipal company of the city of Jena KIJ (NN 2021ab). KIJ operates and renovates around 400 municipal buildings in the city of Jena. The BIM authoring tool Autodesk Revit and the CAFM system SPARTACUS Facility Management are used as software solutions to support the corresponding tasks. Both systems are connected via an integration module. Almost all buildings are available as Revit models, the alphanumeric data is maintained in the CAFM system.

9.2.2 BIM-CAFM Integration

Based on the linking of both systems, not only the space data but also the technical asset data (e.g. HVAC) from the BIM model can easily be transferred from the planning phase into building operations. In this way, fire damperrs, ventilation systems, automatic doors, etc. as well as all information required for RE operations are transferred with a single click. Based on this, all processes related to technical support of the MEP systems can be mapped and executed in the CAFM system.

The BIM integration module then allows easy updating of the information from the operational phase (depicted in SPARTACUS) into the BIM model. This way, the extended information from the CAFM system can be visualized in the BIM model. For example, the color coding of technical assets, whose maintenance dates are overdue or about to expire, is possible in the BIM model. Likewise, other FM process data can be visualized in the BIM model. For example, buildings and rooms with expired leases can be highlighted in color, as well as energy-intensive objects.

At KIJ, MEP data often need to betransferred from the CAFM system to the BIM model. This situation is due to the fact that KIJ has been working with the CAFM system for longer than with the BIM model. The link allows the MEP components, including their properties, to be addressed from the CAFM system in the BIM model. On the other hand, MEP components can be referenced from the BIM model in the CAFM system. The MEP components have the same identification number both in the CAFM system and in the BIM model.

9.2.3 Result

The integration module controls the bidirectional communication between the BIM model and the CAFM system. This way, tasks are always carried out in that software system in which they can be processed most sensibly and effectively. This creates a symbiosis of building planning and management processes from the integration of the two worlds BIM and CAFM.

9.3 Axel Springer New Building in Berlin

9.3.1 The Project

In October 2016, the construction of the Axel Springer new building in Berlin started officially. In the meantime, a modern publishing house has been built on the approximately 10,000 m^2 large Lindenpark site, which was officially handed over to its users after four years of construction on 06.10.2020. With an office space of 52,200 m^2 it offers space for 3500 employees (cf. Fig. 9.1).

The Dutch architectural office "Office for Metropolitan Architecture" (OMA) has created a working environment through its design (cf. Fig. 9.1, left) that promotes both concentration and lively collaboration. This resulted in a spectacular building, which is also intended to lead the media company into its digital age in terms of its architectural claim.

The focus on digitalization was already evident in the first conceptual phase. In this context, it was agreed between the client (Axel Springer SE), the architect (OMA) and the general contractor (Züblin) to plan, build and handover the building to the operational phase using the BIM method. For Springer, it was important to pursue a sustainable BIM approach that takes the operation of the building into account from the very beginning.

The goal is to create a consistent building model—from the first planning phase to the operation of the building. The model was successively enriched with more and more information during this period. This created the basis for an intelligent building that brings together the real and the digital world—the Digital Twin.

9.3.2 BIM Structure and System Environment in the Project

Due to the complex architecture, it was clear to OMA to plan the entire building in 3D. In addition, numerous project participants had to be coordinated, controlled and a large

Fig. 9.1 Axel Springer new building Berlin, design by Rem Koolhaas (Office for Metropolitan Architecture) (left), completed building (right)

amount of information to be exchanged without loss. In order to master this complexity, it was decided to realize the project according to the BIM method.

Since architects, planners and later the general contractor use different modeling tools, it was decided to carry out the project in the Open-BIM process. Data exchange between the parties took place on the basis of the IFC format. For data exchange and coordination in the planning and execution phase, the online platform THINK PROJECT was used.

With the BIM implementation, the client pursued a sustainable and continuous approach to data generation and data maintenance. For this reason, it was very important to the client that data transfer or data integration into building operation runs smoothly.

To support the various FM processes, Axel Springer uses the CAFM system pit—FM. As part of the BIM project, the goal was that the as-built model in the original Revit format is transferred to the CAFM software at the end of the project. From the Revit model, the relevant alphanumeric metadata (manufacturer, type, year of construction, maintenance interval, etc.) for the different BIM objects (room, technical asset) were transferred to the CAFM software pit—FM and are available there for the FM processes. The visualization of the geometry is carried out in Revit. For a smooth data exchange between Revit and pit—FM, both systems are connected via a bidirectional interface.

In terms of sustainable data management, it is important to maintain the high level of data quality throughout the lifecycle of the property. Through the integration of additional technologies and systems such as ERP and BMS, a digital twin is gradually created.

In order for the integration of the as-built model to run smoothly at the end of the project, several data transfer points (data drops) were defined in the project. At these specified times, initial model data for the CAFM system were made available.

BIM management was carried out by the client with the support of an external consultant. The overall coordination of the individual trades was assumed by the general contractor. Here Züblin was responsible for the quality of the overall and coordination model. For quality assurance (collision, plausibility checks, etc.) in the project, Züblin used, among other things, Solibri and Navisworks. Furthermore, Züblin was responsible for creating and updating the BIM execution plan (BEP) and its compliance by the project participants.

At the end of the project, among other things, Züblin handed over the as-built model in Revit format to the operator (cf. Fig. 9.2).

9.3.3 BIM Requirements

The basis for an efficient and accurate data exchange between the participants in a BIM project is a detailed requirements definition.

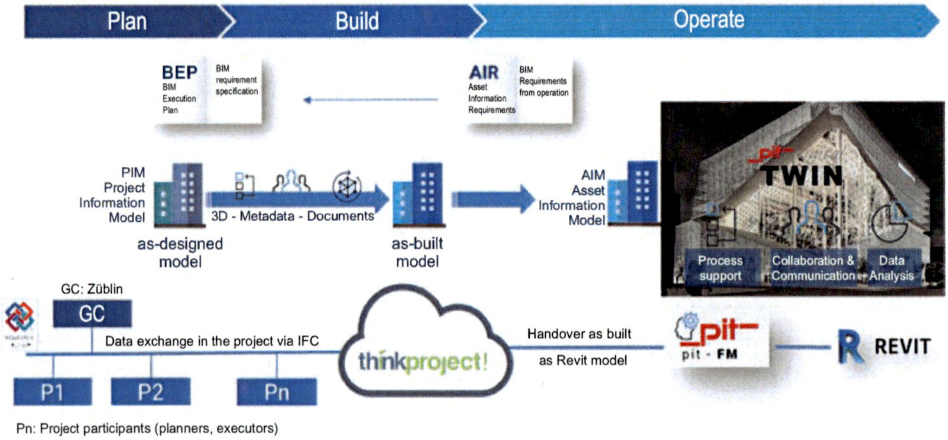

Fig. 9.2 General data and system environment (CDE) in the BIM project at the Axel Springer new building

The requirements for the respective BIM discipline models, their content and level of detail with regard to geometry (LOD) and information (LOI), as well as the processes for data exchange were regulated in the project in the BIM execution plan (BEP).

Specifications for data exchange with a focus on FM or for data integration and transfer to the CAFM system pit—FM were described in the project in a separate document, the Asset Information Requirements (AIR). This document is an annex to the BEP and was binding for the participants.

9.3.4 BIM in Facility Management at Axel Springer

To support the FM processes, Axel Springer uses the CAFM system pit—FM, where geometric and alphanumeric data as well as documents (digital building documentation) are managed, updated and enriched with additional data (cf. Fig. 9.3).

To ensure a smooth and accurate transition from planning and construction to operation, corresponding specifications for FM have been defined analogously to the modeling rules of the planning phase. These specifications are based on the data model of the CAFM system. pit—FM relies on established standards such as DIN 276, DIN 277, IFC and CAFM-Connect, so that integration into the project was no problem. These requirements were already delivered by the client as AIR in an early project phase.

In particular, the following specifications were included in the AIR document:

- Requirements for the unambiguous identification of buildings, floors, rooms and the identification of architectural objects such as doors and technical assets such as heat-

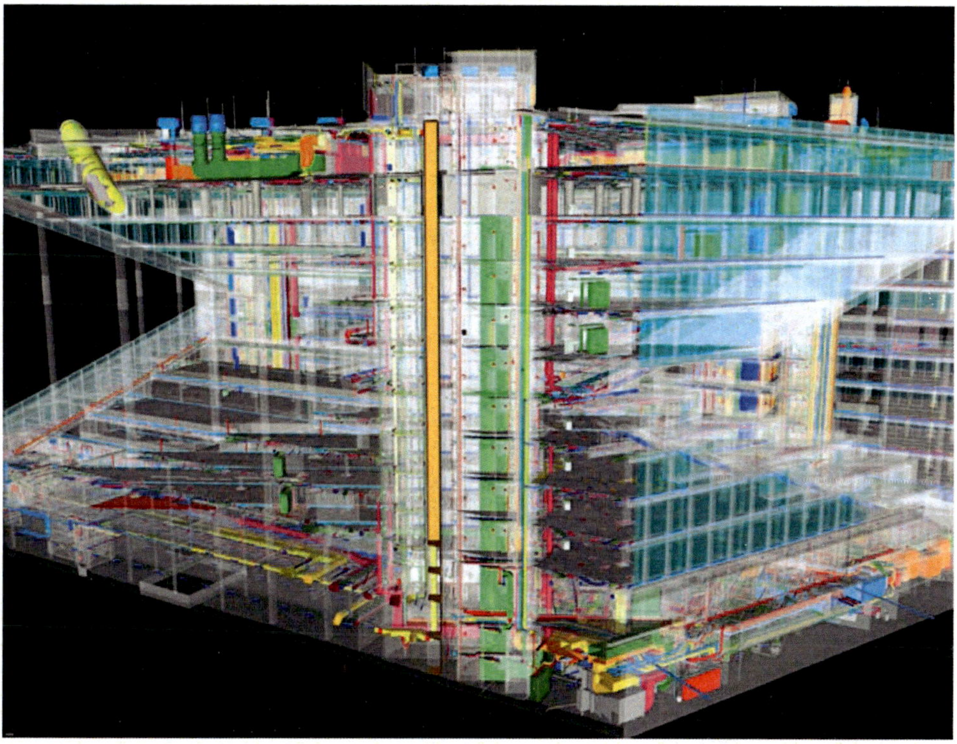

Fig. 9.3 Complete BIM model

ing and ventilation using a consistent asset identification key. Such a key is essential for the unique identification of the modeled objects.

- Definition of a parameter list for the FM model. The list precisely specified which properties (parameters) for the individual objects (door, fire damper, fan, etc.) are required from an FM perspective and when. In order to ensure a smooth transfer or integration into the data model of pit—FM, the objects to be labeled were additionally classified according to OmniClass.
- Description of the process for integrating the Revit model into pit—FM. The transfer of the FM-relevant, alphanumeric object information from the Revit model into pit—FM (pit—FM is the leading system for maintaining the data during operations) was planned here. The geometric integration was achieved via a bidirectional interface between Revit and pit—FM. The 3D geometry was visualized in Revit. In both cases, a mapping rule had to be created once, in which categories and parameters from Revit were mapped to classes and attributes in pit—FM (cf. Fig. 9.4). The use of the consistent classification standard OmniClass made this process possible to a large extent.

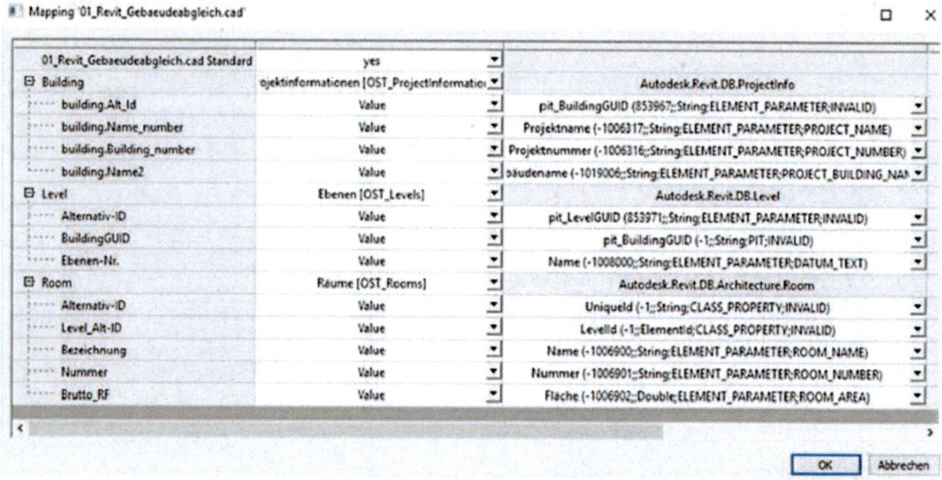

Fig. 9.4 pit mapping rule in Revit

- Furthermore, data drops were defined in the AIR up to the final transfer of the as-built model. Thus, a valid data model was available in pit—FM at the end of the project and building operation could start without delay. The complete as-built FM model including all revision documents was transferred according to the AIR approx. 3–4 months after completion. Metadata from the Revit model, required in pit—FM, for example, for maintenance tasks, were already available to building operation before completion of the building.

9.3.5 Summary

Thanks to the consequent use of the BIM method, the project was completed as planned and handed over to the users on schedule. The seamless transfer or integration of BIM data into building operations and into the Revit and pit-FM systems provided for this purpose took place smoothly and without migration efforts. Those involved in the project therefore describe the project as a successful and sustainable BIM project that was implemented across all service phases right through to building operations.

9.4 Museum of Natural History Berlin

9.4.1 Goals of the Project

The Museum of Natural History Berlin (MfN) is one of the largest integrated research institutions in the field of biological and geological evolution and biodiversity, with more

Fig. 9.5 Future plan of the Museum of Natural History Berlin

than 30 million collection objects, both nationally and internationally. Its mission and vision is combining an excellent research museum with an innovative communication center for scientists and more than 700,000 visitors per year, in order to shape the scientific and societal dialog about the future of the earth on a broad basis.

In order to achieve these goals, the MfN has developed a strategic future planning together with its partners (see Fig. 9.5). Key aspects of this planning are to make comprehensive parts of the collection accessible to visitors, to drive the digitalization of the collection, to test new forms of participation involving society, and to strengthen top research and science communication at the same time. However, in order to implement these future visions, the consequences of the destruction during World War II and a massive need for redevelopment must first be addressed. This includes not only the elimination of structural deficits, but also the modernization of the technical infrastructure.

9.4.2 Starting Situation

The MfN has a current area of 38,203 m^2 at the site of Invalidenstraße 42/43 and has plausibly estimated a need for usable space of approximately 63,000 m^2 for future plans (NN 2021s). This comprehensive renovation and development project has been planned in a sequence of construction phases in order to maintain both research and museum operations.

In the first construction phase, important exhibition halls on the ground floor were renovated and reopened to the public in a first phase from 2004 to 2007. In a second construction phase, from 2006 to 2010, a first step towards the vision of an open and integrated research center could be implemented. With the reconstruction of the east wing, which was almost completely destroyed in World War II, new storage facilities were created in addition to improved storage possibilities, and a new home for the scientific wet collection was created on the ground floor, which is accessible to both researchers and visitors. The second construction phase continues this concept and opens up more areas of the collection to visitors, but also creates space for workstations. In this way, both employees and visiting researchers are provided with workstations in close proximity of the research collection.

A special feature here is the integration of modern, energy-efficient building technology and sustainable air conditioning strategies, such as the use of moisture-regulating properties of clay plaster in combination with an integrated heating and cooling circuit system.

With the third construction phase, research infrastructure and science communication are addressed. In this context, additional, contemporary and flexible workstations are being created for an increasing number of technical and scientific employees as well as for administration. For the third construction phase, which was planned in 2019, the application of the BIM method was agreed for the first time.

In essence, MfN pursues BIM with the aim of increasing planning quality as well as consistent data management, but also better communication and coordination between operator and users on the one hand and participants in the planning phase on the other hand. The required consistent data management is to result in the transfer of as-built models at the end, which are to be used for subsequent new construction and renovation measures as well as for operation in FM. In the commissioned EIR, for example, digital user coordination, a model-based room book and a digital acceptance and fault management were agreed during planning.

As can be seen from the overview of the collaboration process (cf. Fig. 9.6), regular project coordination meetings were held in this case, in which Open BIM (cf. Sect. 3.4.4) was applied. In this case IFC models using IFC 2x3 were exchanged. The BCF format is used for issue tracking. The project space/CDE BiG® is used by the architects Müller Reimann, who have taken over the project as general planner and BIM manager. The software poolarServer is implemented for archiving. The role of the BIM information manager is occupied by MfN. The modeling is carried out mainly with the BIM authoring tools ArchiCAD and Revit in their discipline-specific variants (MEP, structure, etc.). For model coordination, the products BIMCollab and Navisworks are used, among others.

At the end of 2021, the model-based design was submitted for approval. In this context, six IFC models in the field of architecture, two models for structural engineering and five models for engineering technology have been created.

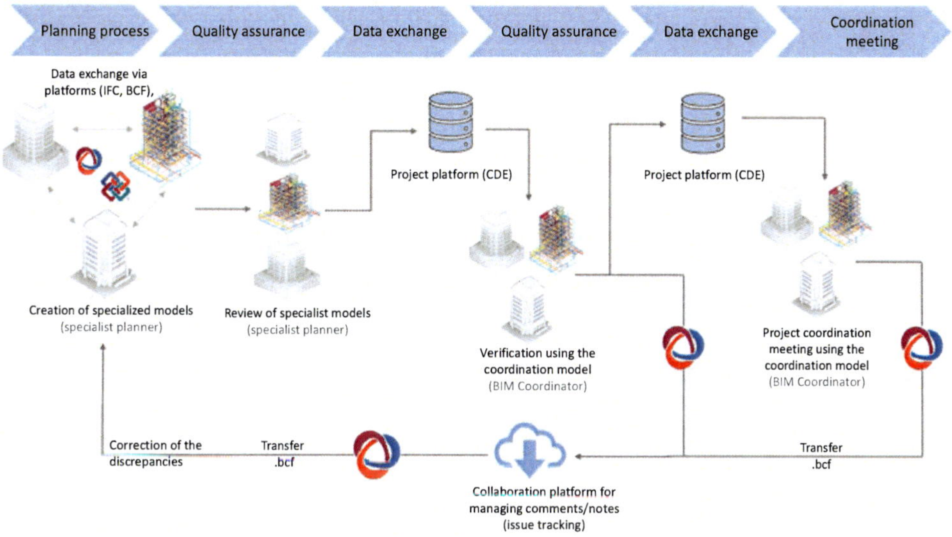

Fig. 9.6 Overview of the collaboration process

9.4.3 Objectives of the Cooperative Research with HTW Berlin

Against the background of this situation and the experiences made during planning for the 3rd construction phase, the MfN decided to build up further BIM know-how with a view to the use phase. This is all the more true as the MfN is also planning to acquire a new CAFM system with the assumption of the MfN's operator responsibility for the property.

With this objective, the MfN and the University of Applied Sciences HTW Berlin have concluded a cooperation agreement. The objective of a first cooperative research project in the field of construction/FM was to examine future information management in FM using BIM in connection with CAFM systems in 2020. This research had to specify the requirements for BIM models to be used in future FM with regard to attribution, classification and reusability for selected BIM applications, as well as to develop procedures for continuous model maintenance.

9.4.4 Approach

In order to achieve the agreed objectives, it was decided to set up a BIM-CAFM demonstrator as a test environment for a CDE. A CAFM system with bidirectional IFC-capable BIM interface is being used for this purpose. The modeling will initially only focus on a part of the building, with the great dinosaur hall on the ground floor being chosen as the starting point. This has a sufficient level of complexity for the examination in terms of both the building and the technical assets.

It quickly became clear that focusing only on as-built models, which have already been detailed according to the FM specifications, will not be sufficient, even if such models are to be expected from the 3rd construction phase in perspective. Rather, it should also be examined how parts of the existing building are to be dealt with in the future, for which no BIM model from a conversion project will be available at first.

Another aspect of the investigation will be the re-use of BIM models in the operational phase for new purposes. Here the exhibition and event concepts from an FM perspective are to be considered in a VR environment. As a result, it was decided to investigate the three BIM model types shown in Fig. 9.7 in the BIM-CAFM Demonstrator.

The *BIM Lite* model (type 1) is limited to those BIM elements that can be captured with low effort or generated semi-automatically, using existing plan documents and additional 3D laser scans. In this scenario, only simple BIM applications such as area calculation are implemented. Most of the alphanumeric information required for operations is entered into the database of a CAFM system. The motto for BIM modeling is: As little modeling as necessary.

The *BIM Complex* model (type 2) contains all BIM objects that are to be transferred from future renovation/modernization projects during construction and planning to FM. This type is used to answer, for example, the question of which BIM objects with corresponding properties should be synchronized with the database objects of the CAFM system. The starting point for this model type are the already available BIM profiles of the CAFM-Connect interface, which can also be used for quality assurance of the BIM model.

The *BIM VR* model (type 3) allows the assesment of the BIM models built for FM in a VR/AR environment. This includes both commercial products and open source solutions such as the community project Blender presented in more detail in Sect. 4.4 with an IFC and VR plugin.

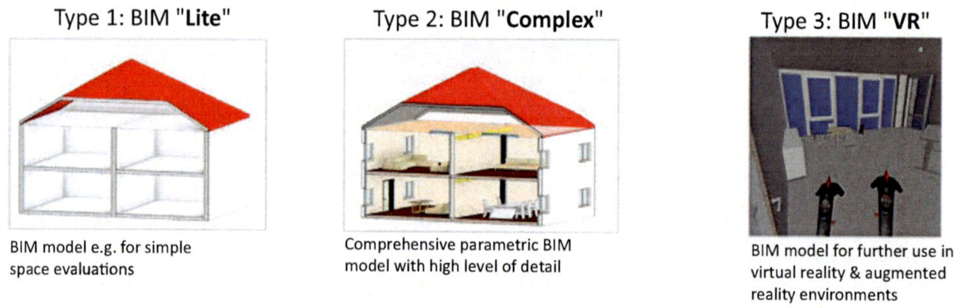

Fig. 9.7 BIM model types for the BIM-CAFM demonstrator

In detail, the procedure comprises the following steps:

- Status determination/assessment of initial data and digital data capture,
- Definition of the BIM use cases,
- Establishment of the IT environment for the BIM-CAFM demonstrator,
- Creation of the BIM models for the three defined BIM types,
- Validation and evaluation,
- Adaptation and extension.

9.4.5 First Results and Expected Benefits of the Feasibility Study of a BIM-CAFM Integration

The first step of the procedure, the status determination and digital capture on site, has already been completed. For this purpose, 3D laser scans were recorded in the exterior and interior of the main building. A Trimble TX8 3D laser scanner with range extender was used for this purpose. In the interior, a total of 61 scan positions were recorded on the ground floor, including the large dinosaur hall. Furthermore, scans were taken of a future archive hall to be opened to the public on the 2nd floor and of its connection via corridors and stairs. In the exterior, the façade of the main building, the forecourt and the path to the rear of the main building were scanned using 28 different scan positions. Unnecessary points in the corresponding 3D point clouds were removeded individually. Subsequently, these clouds were registered in a total point cloud and segmented for easier processing later on (cf. Fig. 9.8).

In step 2 of the procedure three BIM use cases were selected as a starting point. However, a later extension is expected in the course of the project.

As an example of the quantity take-off based on the BIM model as the first use case, the determination of cleaning areas is considered. For this purpose, not only the artifacts in the exhibition hall are examined, but also an adjoining room with glass showcases. In addition to findings for space management, the aim of this use case is also the communication of cleaning concepts and their visualization.

The second use case is chosen from the field of maintenance, in which the lighting elements and the MEP objects contained in the great dinosaur hall are considered. The accessibility of lighting fixtures for the change of illuminants is examined first. In perspective, the use of BIM lighting elements in an VR environment for assessing illumination scenarios to suppport alternative exhibition concepts is of interest.

The third use case looks at navigation, e.g. under pandemic conditions or in the event of an evacuation. Insights are expected from this use case with a view to creating fire department route maps or priority plans for securing artifacts from fire. A further use is planned as part of indoor navigation.

Fig. 9.8 3D point cloud of the large dinosaur hall and the exterior of the building

Currently, work is being carried out on the third step, in which a first test environment is set up using the CAFM system pit—FM. Here the IFC Builder ensures the synchronization and testing of the first version of the BIM Lite model.

9.5 ProSiebenSat.1—Mediapark Unterföhring

ProSiebenSat.1 combines leading entertainment brands with a strong dating & video and commerce & ventures portfolio under one roof. With this setup, the company is continuously driving its diversification from its own resources. In the entertainment sector, they offer the best entertainment—whenever, wherever and on any device, whether with lighthouse formats like "The Masked Singer" or successful in-house productions like "Germany's next Topmodel—by Heidi Klum". With 15 free and pay TV channels, they can reach over 45 million TV households in Germany, Austria and Switzerland. In addition, around 33 million unique users use the online offerings marketed by ProSiebenSat.1 every month.

At the same time, they use their expertise in building brands for two other business areas: With ParshipMeet Group, they have created a leading global player in the dating market, which will significantly support their future growth. As part of the investment and commerce activities, ProSiebenSat.1 builds digital consumer brands such as flaconi, Jochen Schweizer mydays or Verivox with their TV reach and advertising power and turn them into market leaders in their respective industries. They are a strong growth partner for digital companies. ProSiebenSat.1 is supported by over 8200 employees who inspire the viewers and customers every day with great passion (NN 2021ao).

In Germany, ProSiebenSat.1 operates together with its majority stakes in more than 28 locations. The focus of the case study is the Unterföhring site, being with 16 buildings and more than 100,000 m^2 of gross floor area (GFA) the largest and most important site.

9.5.1 The Project

ProSiebenSat.1 operates, distributed over the Unterföhring media park, 16 structurally divers buildings. The buildings differ greatly in size, construction, year of construction, documentation and usage rights (ownership/leasing). So the largest building has a floor area of 2500 m^2, while the smallest building, in which there are workplaces, only measures 200 m^2. In addition to areas for studio operations and production, 3500 office workplaces have been set up in the existing buildings.

ProSiebenSat.1 has decided to start modeling its existing buildings as early as 2018 in order to gradually build up competence in the field of BIM and BIM2FM. The goal was and is that all existing buildings are documented in a comparable data quality and can be operated with the help of as few data sources as possible based on uniform information and processes.

For the inventory and modeling, TÜV SÜD Advimo GmbH has created BIM employer information requirements (EIR) and a BIM planning manual. The inventory is carried out by the BPS Group on this basis. The BIM planning manual serves as a guide for the inventory data acquisition, including re-modeling of the existing buildings based on 2D templates and on-site capturing of the architecture. In a next step, the MEP data capture takes place. The quality of the inventory is secured by the BIM management of TÜV SÜD Advimo.

An upcoming conversion (strengthening of fire protection) is pending and is also to be realized with the BIM method. These projects are implemented, inter alia, by an internal planning team, including the project management, of approximately six employees, external architects and specialist planners, external execution companies, external BIM consultants and BIM managers for quality assurance (TÜV SÜD Advimo), external re-modelers and data acquisition companies (BPS Group) as well as an external CAFM vendor (eTASK).

The BIM requirements of the client have a strong focus on the operational phase, in which the CAFM system is used, e.g. for the administration of the spaces, workplaces, parking spaces, locking systems and for cleaning management.

For the collaboration of all parties in the phases of planning, construction and operation, the CDE from Autodesk (BIM360 Design) is used and coupled with eTASK's CAFM system via a Revit plugin. The BIM EIR for the inventory were coordinated with this target IT environment ensuring smooth data transfer. The EIR describe in detail how existing (or newly captured) 2D data is implemented in 3D BIM models using the BIM authoring tool Revit. These models are then utilized as BIM models for operation in the CAFM system.

9.5.2 The Aim of Using BIM

The use of BIM methodology is intended to ensure the completeness of information in the BIM model for operation. Furthermore, new methods such as the automation of processes, e.g. door numbering by Dynamo scripts and application to the BIM model, are to simplify daily work.

In the past, existing documents (planning basis) were completely changed by planners during conversions and were no longer CAFM-compatible after return. In the future, external planners will be connected to the CDE and will be able to use real-time modeling capabilities and dynamic permission concepts for external planners via BIM360 Design. This will eliminate the need for exchange and upload of files in future as well as the manual quality assurance required for this.

The coordination between internal and external planners was time-consuming in the past and can now be automatically regulated on the basis of the permission concept at the element level. The BIM models therefore no longer leave the organization and the central model can be synchronized with the CAFM system at any time.

9.5.3 Approach

For implementing the BIM methodology in projects where existing buildings are remodeled, but also in projects where occuring changes must be mapped, the EIR specifically developed for each project category form the basis.

The PAS 1192 standards (NN 2014b) and in particular the modeling guidelines and basic templates for the contract definition for external planners and their subcontractors form the basis for the creation and revision of BIM models (cf. Fig. 9.9). These standards have been adapted and coordinated with the participants during BIM implementation and are meanwhile replaced by the ISO 19650 standards (NN 2018c).

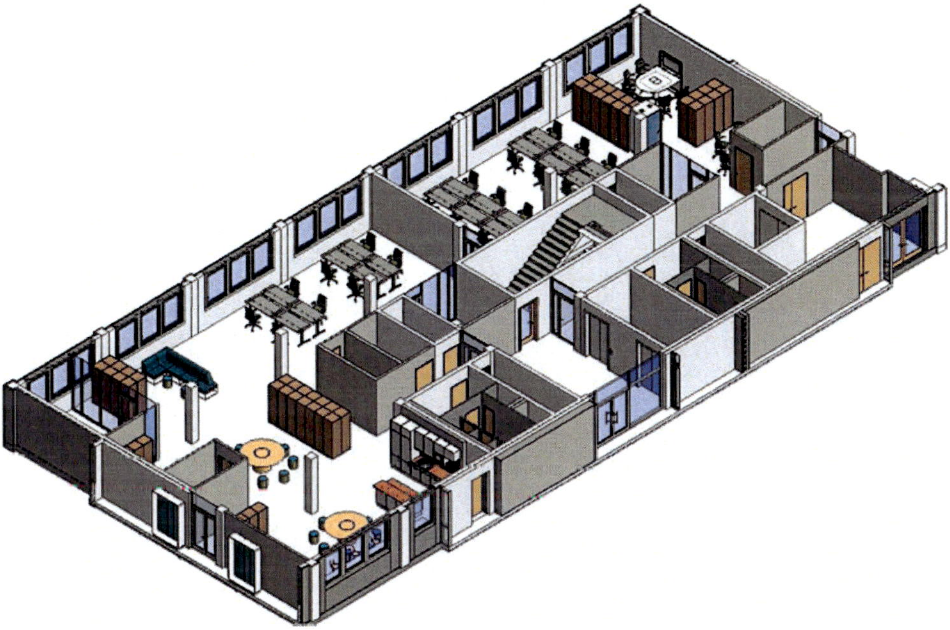

Fig. 9.9 BIM model—floor including furniture

The following documents were created with a focus on room book applications and asset usage:

- EIR for as-built models as well as for the creation and use of Revit as-built models in eTASK with uniform minimum standards (in PDF and DOC format) and focus on the native Revit format (RVT),
- EIR for awarding of external planning services (refurbishment of existing buildings) for architecture, building technology and structural engineering with cross-object, uniform minimum standards (in PDF and DOC format) and focus on the native Revit format (RVT),
- A template of the BIM execution plan (BEP in DOC format) for awarding of external planning services (refurbishment of existing buildings),
- Definition of minimum BIM requirements (in PDF and DOC format) for a Revit template readable into CAFM as well as for shared parameter files with specifications, which external and internal model managers must observe, in particular with regard to
  - The template structure (project browser incl. views, sections, floor plans, etc.),
  - Shared parameters,
  - Object families,
  - Model coordination and model coordinates, and
  - The creation of an overall model and its delivery quality (central model), see Fig. 9.10).

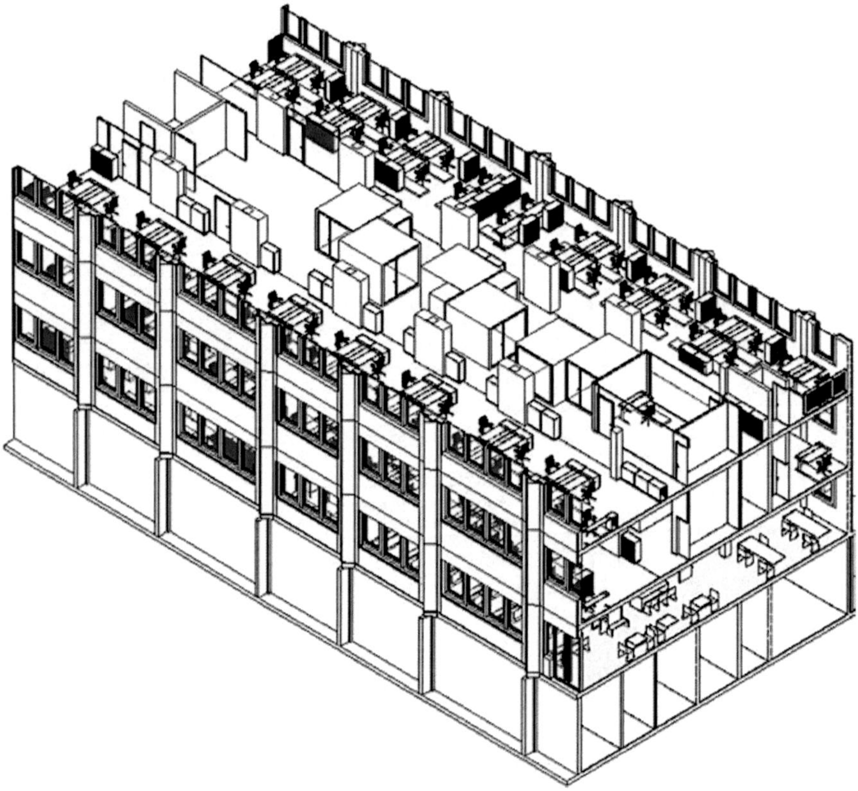

Fig. 9.10 Multi-story BIM model

The BIM execution plan (BEP) serves as a model for updating by future contractors of ProSiebenSat.1. The BEP template was created with the aim of implementing the framework conditions already defined in the EIR. Subsequently, the definition and deepening in the collaboration process was carried out by all participants—client, BIM manager, planner and specialist planner.

The successful implementation of the application cases specified in the EIR is delimited by the following parameters in the BEP template: Who delivers what? In what degree of detail? At what time? In what format?

Together with the participants, the setup of a structured document and model management for project collaboration, a common data environment (CDE), based on the PAS 1192 standards, in particular PAS 1192-2: 2013 and PAS 1192-3: 2013, with data privacy and data security criteria according to PAS 1192-5: 2015 was carried out. In addition, the current processes were recorded during planning and target processes were defined. They include best practices for collaboration with BIM models and are specifically adapted to the organizational and general process-oriented structure at ProSiebenSat.1. The roles and rights for editing BIM models, which should later be implemented in the document management and organizationally in the processes, were defined.

In order to enable the individual external planners not to use different Revit templates and to meet the objectives and requirements of the EIR, the external planners were trained and equipped with common shared parameter definitions allowing them to correctly implement the requirements.

The internal employees were trained in parallel on:

- Bidding strategy,
- Project platform (CDE or project portal),
- Software and best practices for setting up modeling projects,
- Roles and responsibilities,
- Data drop strategy, schedule and delivery requirements,
- Modeling guidelines,
- LOI and LOG implementation,
- Coordination between planners, BIM manager, GC or PM and client,
- Implementation of project structure plan in BIM use cases,
- Implementation of standards and
- Collision checking.

Workplace planning and changes to floor plans are carried out on this basis in Revit and transferred as IFC files with the relevant information into the CAFM system eTASK.

As part of BIM implementation and conversion of processes, weekly jour-fixe meetings are held with the internal and external planners, the inventory data collectors and modelers as well as the CAFM implementer to coordinate the *lessons learned*.

TÜV SÜD Advimo checks the individual discipline models of the planners and modelers for the modeling rules agreed in the EIR's, LOD's (Level of Details) and LOI's (Level of Information) so that the BIM use cases can be implemented successfully. In addition, the collision checks are examined for their plausibility and the structure of the test queries. The results of the model checks are handed over to the client and the planners and modelers in the form of a test report and are discussed together.

9.5.4 Conclusion

A major challenge in the implementation of BIM was the conversion of the existing manual and established processes. Therefore, it was necessary to integrate the change management concept in addition to the expert support of all parties involved.

The extension of the standard IFC data transfer by parameters significant for operation was another challenge in order to avoid manual post-processing of missing or incomplete information.

The Revit specifications must be adhered to consistently and the parties must not deviate from them during the course of the overall process. There must be no break in the flow of information—from re-capturing or -modeling to CAFM data import—otherwise the data stocks will inevitably diverge from each other in case of changes.

The BIM model with a clear maintenance responsibility must be the central and leading data basis on which the CAFM system and the operational processes can then be based.

9.6 BASF in Ludwigshafen

9.6.1 The Project

BASF is the world's leading chemical company. Worldwide, more than 110,000 employees work to make customers successful in almost all countries of the world. With high-quality chemical products and intelligent solutions, BASF contributes to finding answers to global challenges such as climate protection, energy efficiency, nutrition and mobility. There are six integrated sites (Verbund sites) and many other production sites worldwide. The world's largest contiguous chemical complex, owned by a single company, is located in Ludwigshafen. On an area of 10 km^2 34,000 employees are employed in development, testing, production and sales of more than 8000 different products.

Of the more than 2000 buildings on the factory premises, more than 400 are cared for by the central real estate unit. As an internal provider of space, the unit ensures that office, laboratory, workshop and storage space is available on the site as required, in a timely and economical manner. The laboratory space amounts to more than 100,000 m^2. The experts regularly ask themselves to what extent the space in older buildings still meets the requirements of the users, needs to be renovated or it is more expedient to replace it by a new building.

In the case of the project presented below, the previously used building and laboratory space was outdated. Detailed analyses of the building substance and the future requirements for laboratory space from the user's perspective as well as economic calculations led to the decision to build a new laboratory.

The building currently under construction (see Fig. 9.11) offers approximately 5500 m^2 of laboratory and cleanroom space. In addition, offices, meeting rooms and social rooms are provided for over 200 employees. The extensive laboratory and ventilation technology is housed on three technical floors.

Digital documents are key not only to planning and construction, but also to building operation.

Due to the complexity and investment volume of this construction project, BASF decided at the beginning of planning in 2017 to use as holistic a building model as pos-

Fig. 9.11 Laboratory building in November 2021

sible for planning and execution. This enabled to erect the building as a whole with the support of the BIM methodology, and to further use relevant data during operations.

9.6.2 BIM Pilot

Since the internal real estate and planning department did not have sufficient BIM planning competence at the time, BASF decided to initiate a "BIM pilot project" with the support of an external partner who already implemented own BIM construction projects from the client's perspective. His tasks included supporting the establishment of an appropriate project structure, conducting laboratory-specific simulations and being responsible for BIM quality management in planning and implementation. In workshops, expectations, goals towards BIM and use cases related to planning, construction and operation were determined. In addition, detailed specifications for BIM planning were defined.

Two points were decisive for a subsequent search for a suitable project planner—on the one hand agreeing on a closed BIM method and on the other hand that the building must be planned with Autodesk Revit. These specifications resulted from the desire to maintain the digital building twin at BASF after completion of the construction and achieving the as-built status. During a call for tenders, a general planner was selected who not only had the competence for constructing a complex laboratory building, but also gained BIM experience in planning comparable laboratories.

"Current BIM models focus on planning and construction. Within the framework of the pilot project, BASF wanted to take a step further. How and to what extent the models can be used for the operational phase was therefore analyzed in parallel with the planning," says Hagen Förster, project operation manager of the new building.

9.6.3 BIM-CAFM Integration

Many processes for controlling the centrally managed real estate at the site are supported in addition to SAP by a CAFM software. Therefore, a main focus was on the integration of the BIM model into the CAFM system. The IFC interface proved to be suitable for transferring complex data into the CAFM system. In order to keep the data set in the CAFM system manageable, filters had to be created when importing data. In a CAFM pilot environment, BIM models were imported at different planning stages, performance tests were carried out and optimization potentials were identified.

A challenge appeared during visualizing walls in the 3D model of the CAFM software. Here, additional adjustments had to be made in the software in order to obtain space-defining elements from a wireframe model. By importing the data directly via the IFC interface, the CAFM software was overloaded with an unnecessary level of detail. This had a strong impact on the performance of the software. Only by using filters, which were used both for the import and later visualization, was the problem able to be fixed. Component-specific features from the BIM model could be directly transferred to the CAFM software (see Fig. 9.12). It has yet to be specified in the coming months how much of this will remain in the software during the operational phase or will be maintained in SAP in the future.

In addition to the integration of the BIM model into the CAFM software with over 4000 floor plans, further sub-projects were necessary from the CAFM perspective:

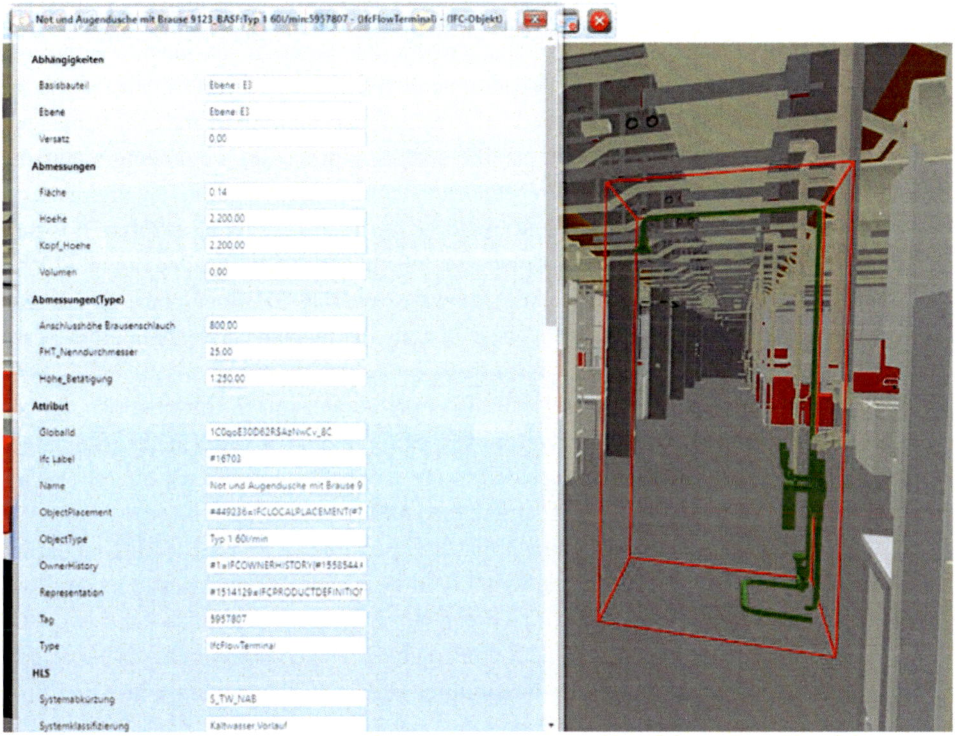

Fig. 9.12 Import of attributes via the IFC interface into the CAFM software

"Today's methods of operating buildings had to be analyzed. The question arose as to how an digital building twin can be economically operated in the same software alongside many other buildings with significantly less information," explains Patrick Holl, responsible for the CAFM software. The focus here was on maintaining comparable operating processes.

Furthermore, a new interface to SAP-PM was created via an existing middleware. This made it possible to access the corresponding SAP information directly from the graphical representation of an object in the CAFM software. Interfaces to technical assets were tested and piloted in order to visualize measurement values directly in the CAFM interface.

The integration of 360-degree images was further developed as an addition to 2D and 3D floor plan information.

9.6.4 Results and Experiences

Working with building models and spatial images required the inclusion of 3D objects. In a first step, beside existing 2D furniture symbols, additional 3D furniture objects were integrated into the software for occupancy planning. Here, too, the objects could not be imported directly from the corresponding graphic databases of the manufacturers with their wealth of detail information, but had to be simplified due to performance problems.

The fact that not all buildings stored in the system have approximately the same depth of information does not prevent comparable operating processes from being maintained across the entire building portfolio. Whether it is worthwhile to completely digitize existing buildings for operation subsequently is to be assessed case by case. Often it is more economical to augment existing floor plans by 360° images to approach a digital twin.

The CAFM software used by BASF allows to overlay 2D floor plans by 3D BIM models and 360° panoramic images. These can also be displayed on mobile devices on site.

Furthermore, not only properties, but also graphic and image information from different points in time can be stored in the software (cf. Fig. 9.13). Thanks to this technology, the real estate department decided to use a demand-driven, cost-effective hybrid approach in the future to document its more than 400 buildings.

The piloted sub-projects have shown that it is possible to transfer a BIM model into a CAFM software and to maintain a digital twin by means of interfaces to existing systems.

The laboratory building is still under construction. The commissioning is expected in 2022. Against this background, the possibilities shown in the piloting must first prove themselves in building operation.

The BIM pilot project has provided insights into how BIM plans for new buildings can be standardized in the future and integrated into the complex system landscape afterwards.

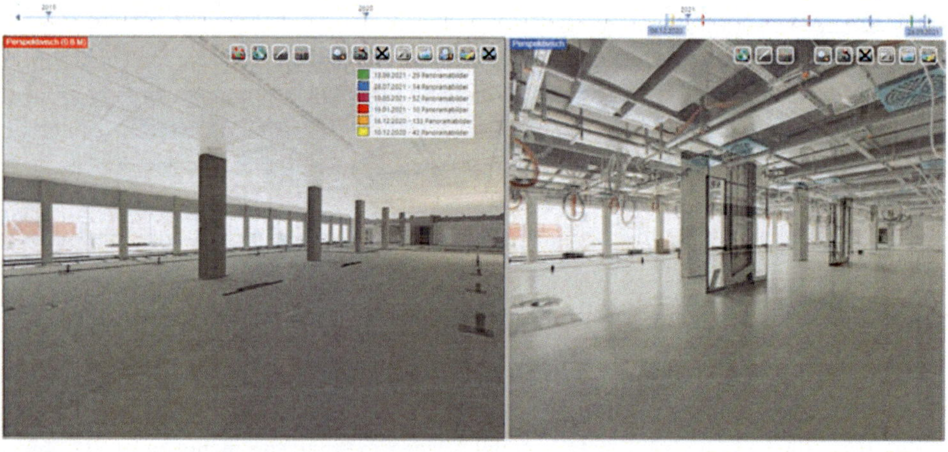

Fig. 9.13 Comparison of different construction states in the CAFM software

9.7 TÜV SÜD @ IBP in Singapore

9.7.1 The Project

TÜV SÜD is a trusted partner for quality, safety and sustainability solutions. For 150 years, the company has created value for customers and partners—with a comprehensive portfolio of services in the areas of testing and certification, auditing and consulting.

Today, TÜV SÜD is represented in more than 1000 locations worldwide with more than 25,000 employees. TÜV SÜD's Singapore office moved to a new integrated laboratory and office building in 2021, housing 600 employees from TÜV SÜD PSB and TÜV SÜD Digital Service Centre of Excellence. The new building is located in the International Business Park (IBP) and covers approximately 18,900 m^2 (see Fig. 9.14). The project volume was approximately 100 million Singapore dollars.

The latest technologies were used to ensure energy efficiency and sustainability in the later operating phase. The office spaces of the new building meet the requirements of the local standard Green Mark Platinum for energy efficiency and sustainability. To increase employee motivation, productivity and well-being, the building also has a fitness center that offers sports programs for employees, workstations in green areas and roof gardens, as well as a connection to the path that connects the public parks in Singapore.

The building was planned and erected in the classical two-phase process from extended rough construction plus building equipment including building technology,

Fig. 9.14 TÜV SÜD @ IBP building in Singapore

interior construction and electricity as well as BMS. TÜV SÜD Advimo used BIM to assert the interests of the owner and user, which is required in Singapore. The current public construction standard of the building authority provides for the use of LOD 300 models for approval planning. TÜV SÜD Advimo extended the local mandatory standard with further, own BIM standards, which concentrate mainly on the optimization of FM processes.

Important applications with a focus on CAPEX and OPEX reduction were set up via internal BIM consulting with process, cost and risk analyses, such as the use of BIM models during planning and construction. For example, maintenance area and movement area models were used for the design and performance optimization of ventilation, air conditioning, (waste) water and electricity. The floor-by-floor construction was also cyclically scanned on site and overlaid with the BIM model to avoid the risk of undetected deviations. Such deviations often lead to ad-hoc changes to the technical equipment or building structure. A CAFM data standard was integrated from the beginning of planning to save the planners the effort of manually attributing spaces and assets. Special TÜV SÜD-own model checkers for WiFi signal optimization and cleanroom pre-certification ensure the quality of the planning.

BIM was used on this project to identify and avoid problems in the planning phases that could so far hardly and only costly be solved ad hoc on site. The BIM model should therefore not only serve as a new form of project documentation, but also support the digital prototyping. Instead of planning conventionally, a combination of virtual construction, simulation and optimization was used to achieve the most cost-effective, risk-reduced and construction- and operation-oriented implementation. In addition to the employer information requirements (EIR) and the BIM execution plan (BEP), the common data environment (CDE) and the as-built-plus-BIM2CAFM data standards for the project were determined by the BIM consultants. After the start of planning, the BIM managers of TÜV Süd Advimo also began develop time-saving and quality-assuring model checkers for cleanroom analysis and the testing of fire protection and escape routes, shading, WiFi signal optimization and maintenance areas for safety-critical assets according to local standards.

9.7.2 Approach

The project with its numerous and sometimes highly complex applications was accomplished using native Autodesk Revit and a modeling guide for model creation, but also for collision checking, BIM objects and the future CAFM use of the models. The local planner and the commissioned general contractor were additionally trained by TÜV SÜD BIM experts and accompanied by the BIM management for high model quality.

To ensure timely and smooth cooperation of client BIM management and the BIM trade coordinators of the contractors, a BIM execution plan (BEP) was developed and updated, and checked by TÜV SÜD Advimo. The BEP is the central document that should be updated on a regular basis and aims to capture and specify all BIM decisions and processes for service providers and for the client. In order to achieve a high model quality for implementating the use cases, the BIM management included, among other things, the following measures:

- Check and optimization of the common parameter files,
- Check and optimization of BIM project templates and families,
- Check and optimization of collision setups including model checks and routines in Navisworks, Dynamo and in the Revit Model Checker,
- Quality and content check of compliance with the BIM standard and modeling guidelines including TÜV SÜD maturity analysis, if provided by the GC.

In preparation for the BIM jours fixes, audits and quality management for the core-and-shell LOD 300 BIM model were carried out during the construction phase. The review and evaluation of the BIM models with regard to the specified BIM requirements in terms of LOD and correct CAD technique (e.g. model orientation, correct creation and application of libraries, closed room polygons, correct connections of piping).

The BIM model check included:

- Checking the quality of the geometric model in terms of true virtual construction quality,
- Checking and adjusting libraries that are generally integrated in construction planning,
- Analysis of building technology and cleanrooms for functionality, buildability, maintenance friendliness and compliance with the BIM models,
- Engineering check of the collision cases and the collision protocols,
- Checking the results of collision detection, e.g. in terms of tolerances, distances and materials,
- Checking the data settings (completeness and quality),
- Checking the technical systems and assets in the model (closed/valid),
- Plausibility checks of the model-based bill of quantities (BoQ),
- COBie integration and COBie completeness checks including supplementing these data standards with additional, object-specific attributes for spaces and assets with a view to the later operational phase,
- Checking the completeness and correctness of the model-based calculations, e.g. for heating and cooling loads and pipe network calculations.

In order to also check the planning-compliant and defect-free implementation during execution, the construction site was regularly scanned from the inside and the outside.

The laser scans were overlaid with the BIM model and thus deviations between the planning and execution could be identified based on AI and displayed subsequently.

It was important above all to interlink all parties according to the BIM methodology pursued and to enable collaborative planning, construction and operation (cf. Fig. 9.15).

The BIM methodology and data base are key to the operation phase. They are part of the future system environment of the TÜV SÜD @ IBP project (cf. Fig. 9.16).

This system landscape aims at implementing the digital twin approach of a connected/autonomous building in reality and testing it for other buildings and customer projects.

The data of the BIM model were imported as master data basis in the operational CAFM system and the client-side control platform. In a first step, this platform is to form the link between the building management system, the rule-based Technical Monitoring System, the Predictive Maintenance Lift Monitoring System and the external service provider's system. Further expansion stages are planned.

9.7.3 Results and Experiences

The BIM methodology and in particular the model tests (clash detection) repeatedly uncovered planning defects which could be rectified in due time before construction and hence avoided costly adjustments during construction.

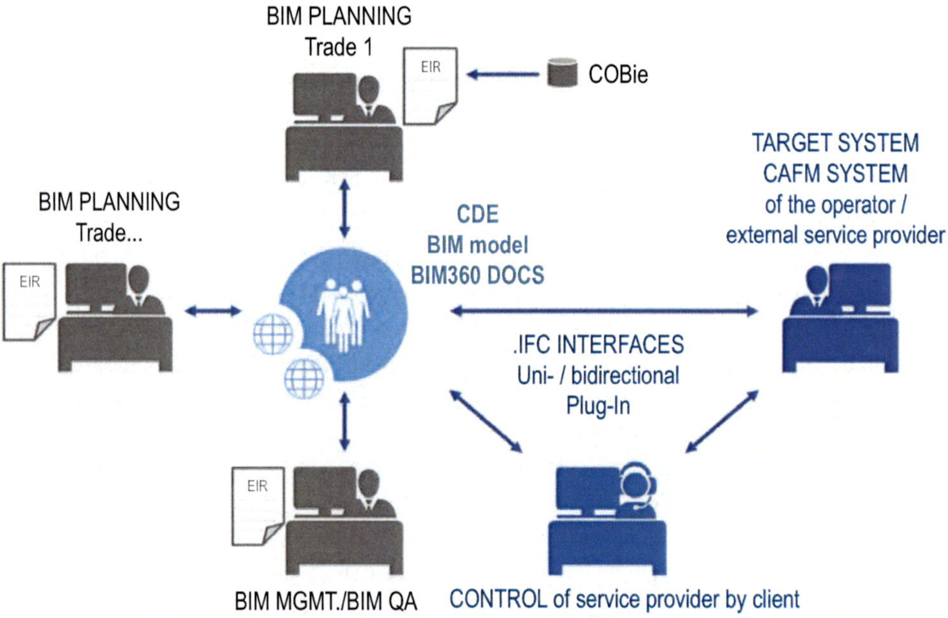

Fig. 9.15 BIM-based collaboration in the project and in the operation phase

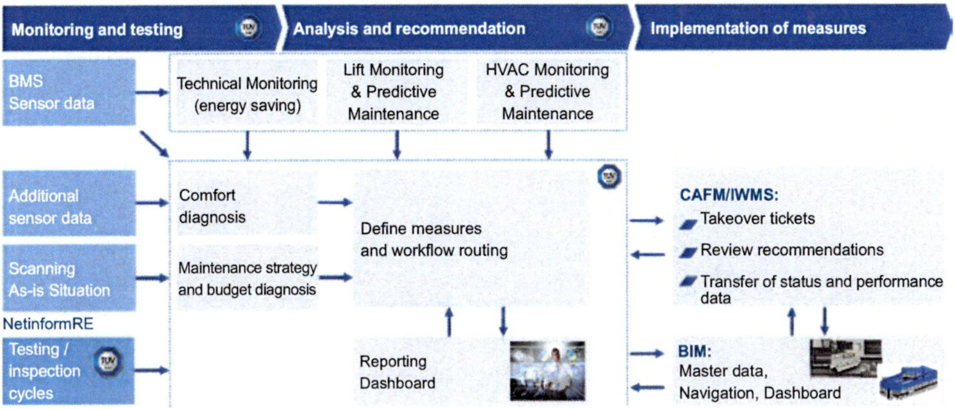

Fig. 9.16 Target system environment of TÜV SÜD @ IBP

This way, for example, the following deficiencies were detected (see Fig. 9.17):

- Component collisions,
- Non-compliance with EIR requirements,
- Missing and inconsistent naming conventions for components and BIM families (information/detail levels in file name, use of underscores, etc.),
- Missing component lists and information such as room, window, wall and door lists.

The application of BIM methodology made it possible to carry out a much more detailed review and evaluation of the planning and optimization approaches. By taking into account the planning/construction-accompanying dependencies in time during the operational phase, the operation of the building could already be optimized during planning. The combination of virtual construction, simulation and optimization contributes to achieving the set objectives in terms of functionality, value retention, compliance with law and sustainability also during operations.

In the operational phase, the following advantages are pursued by TÜV SÜD @ IBP by combining BIM methodology and an extended digital twin approach of the connected/autonomous building:

- Demand-driven reduction of maintenance cycles and costs,
- Reduction of the effort for searching and capturing current and correct information for external service controlling,
- Reduction of the effort for creating and maintaining reports including the evaluation, e.g. of financial risks and the measures to be taken,
- Faster determination of needed actions and faster implementation of measures,
- Reduction of manual communication of repair triggers (autonomous building),
- Avoidance of interruptions of the core business and sales losses,

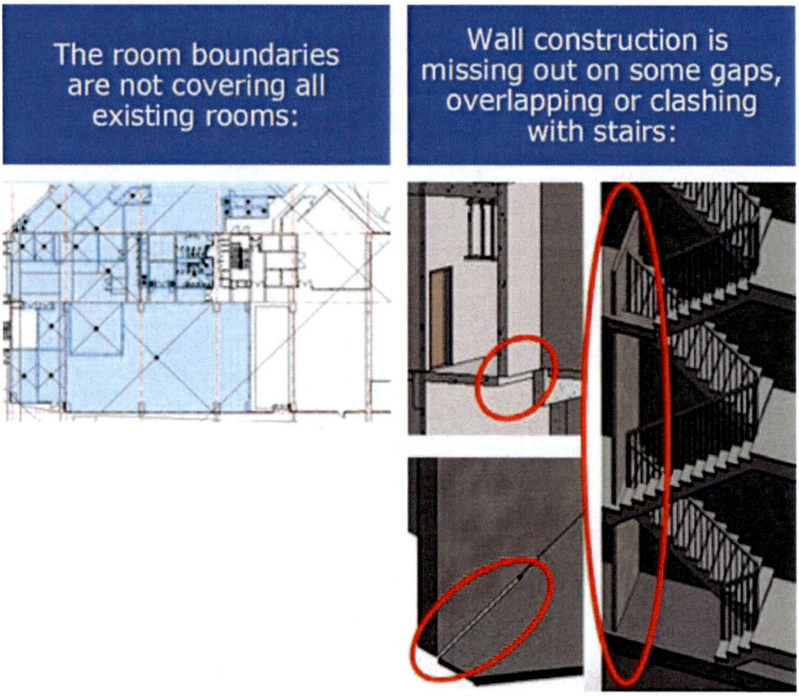

Fig. 9.17 Examples of deficiencies detected by collision test (clash detection)

- Reduction of energy costs through better adaptation of BMS control,
- Increase in efficiency of work preparation due to current and complete technical data (mobile available),
- Acceleration of the search and identification of technical assets by BIM-based navigation and QR/RFID tags,
- Enabling faster and easier just-in-time work documentation and reporting through mobile devices.

9.8 Country Park III in Moscow

9.8.1 The Project

The BPS Group, founded in 1992 as an IT company, is one of the leading providers of BIM technology in all phases of the lifecycle of real estate. The group brings together architects, engineers and IT specialists in a unique conglomerate.

The focus is on the creation of building information models (BIM) and the development of solutions for effective lifecycle management, such as 4D/5D and industrial IoT solutions.

By combining the functions of investor, developer and operator of its own 100,000 m^2 commercial real estate, the BPS Group has extensive practical experience in the development and operation of large real estate portfolios.

Country Park III is the BPS Group's own development project, which was commissioned in 2014. The building (see Fig. 9.18) has a total floor area of 44,300 m^2 and a total rental area of 27,800 m^2. Country Park III comprises a 22-storey A-class office tower, a medical center and 256 underground parking spaces. Leading international companies such as AMD, BMW, Volvo and Kärcher have chosen Country Park to rent office space.

Since Country Park III belongs to the BPS Group, the building has become an ideal location for testing the company's own digital products.

The goal of the project was to use a BIM model throughout the lifecycle of the building.

9.8.2 Approach

The operations department was handed over a BIM model that corresponds completely to the built object (cf. Fig. 9.19). It is the global mission of BPS to organize the teams in

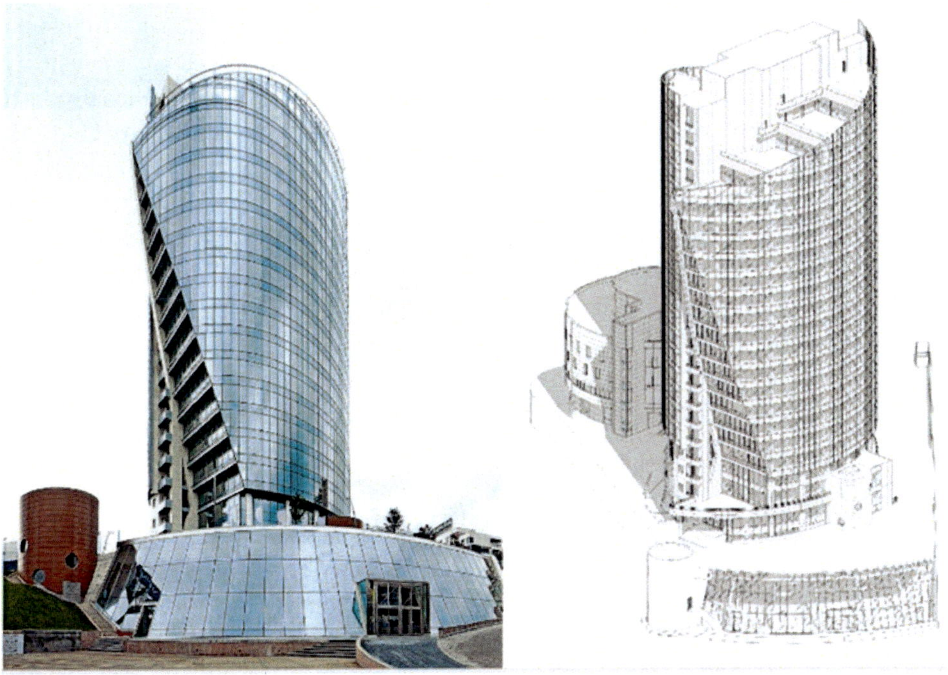

Abb. 9.18 Country Park III

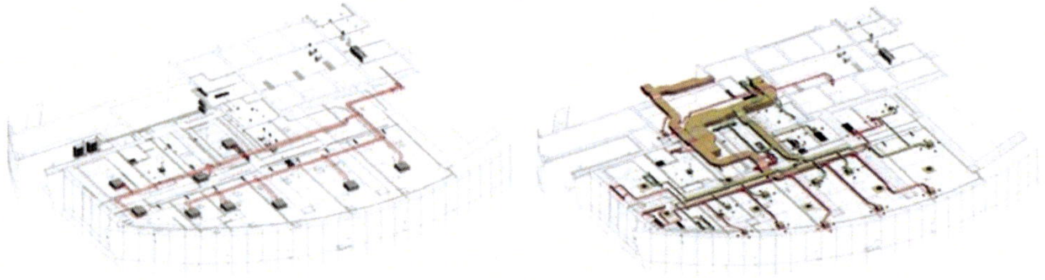

Fig. 9.19 Coordination of the MEP trades, Revit model, Country Park III

all phases of the real estate lifecycle as a single automated process with uniform standards and digital assets.

When defining the requirements for the use of BIM, the focus on the operational phase was decisive in line with the principles of sustainability.

The planning method Building Information Modeling was the basis for all processes in this project:

- Planning was carried out in Revit with the latest software for planning and structural analysis.
- The construction process was supported by the parametric models. The project included 11 different Revit models and more than 66,000 parametric model elements. Suitable attribution made the Revit models the central information source for building management. The responsible project manager had a practical database available with the 3D model, in which the individual building elements were additionally linked with information on assembly or commissioning. This combines all the information for optimal real estate management. The time-consuming search for corresponding information in overloaded shelves with folders belongs to the past.
- The BPS Group also includes a company specialized in facility management, which uses the model as a database for building operations.
- QR codes are placed in all Countrypark rooms. These QR codes are scanned by means of a tablet application using the camera and allow the technical specialist for operations and maintenance (facility manager) to easily locate a reported problem and to efficiently carry out the necessary operation processes.

9.8.3 Integration of BIM and Industrial IoT Technologies with HiPerWare

During the operation phase, the BIM model of the assets was supplemented with additional information: operating modes of the technical equipment, operating instructions, materials used in processing, expiration dates, etc. The process of creating the operational BIM model is not yet completed, in the future a digital twin with a high level of detail will emerge.

The BIM model gives an idea of the "anatomy" of the building, but does not provide any information about the "feeling" of the building. In order to create a proactive management system that recognizes potential "diseases" at the first symptoms, the building and its performance must be examined in the same way as modern medical devices examine human health.

For this purpose, BPS Group has developed the HiPerWare platform, which is based on Big Data, machine learning/AI and industrial IoT technologies. This was used for the first time in Country Park III to bring the BIM model to life. The animated virtual model simulates physical processes that take place in a real building, it is equipped with artificial intelligence and can remember, understand, analyze and optimize these processes (cf. Fig. 9.20).

The first task of the HiPerWare platform is the massive collection of big data from the operation of the technical assets in the Country Park. This is done in real time and is very efficient—with hardly any investment, without interruption of operation or production, without intervention in the systems and without programming or construction work.

The information is synchronized with the cloud storage. The collected and clearly structured big data (billions of variable values) are ready for processing by AI and machine learning algorithms and form the basis for the use of a self-learning neural network.

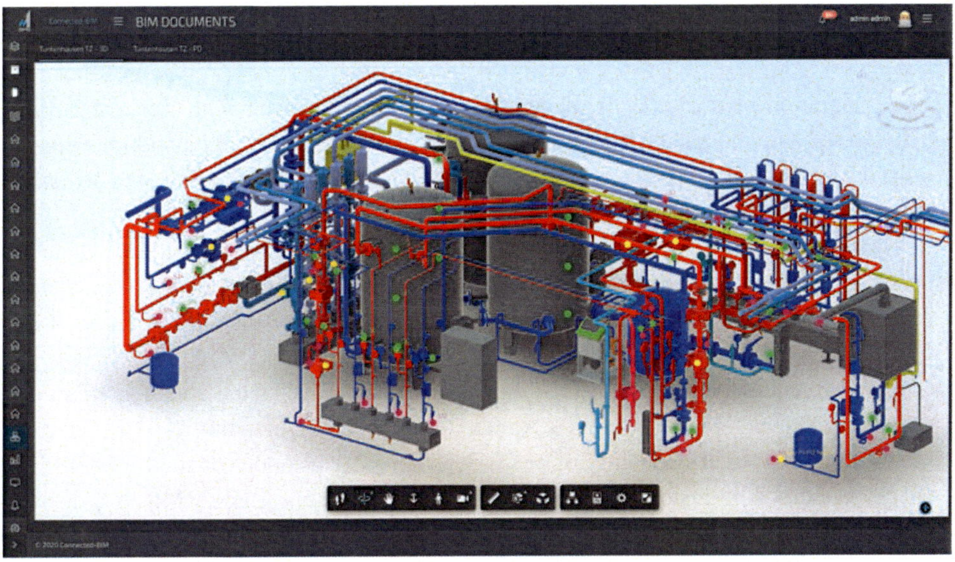

Fig. 9.20 BIM model with live data of the heating system

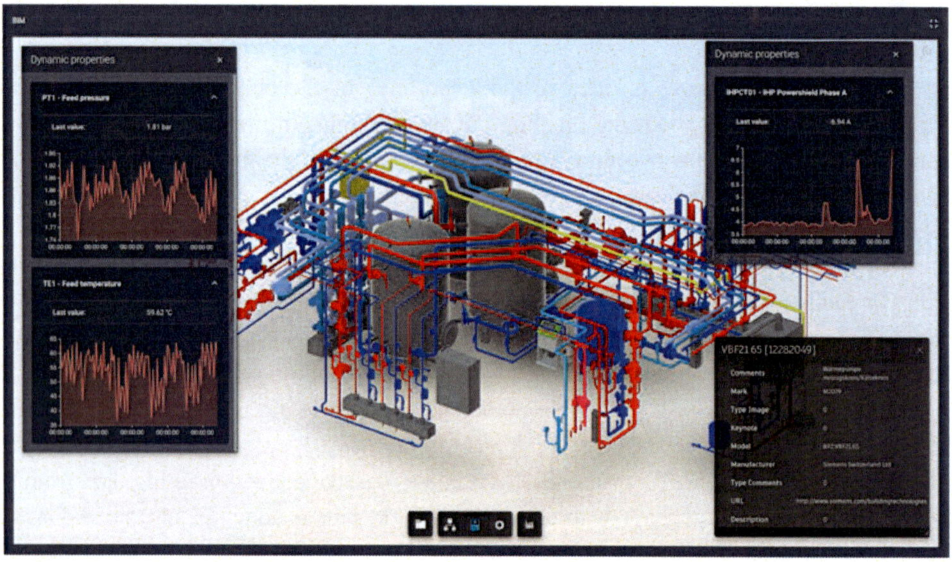

Fig. 9.21 HiPerWare dashboard

Big data is combined with a huge amount of information about the operation of the technical equipment as a whole, such as energy distribution balance, dynamics and values achieved of process parameters, lighting, temperature and air quality, as well as operating modes of devices (cf. Fig. 9.21).

9.8.4 Results and Experiences

The integration of a digital twin, which is fully synchronized with the real building, helps to see and understand the real physical process over time and its relationships with other processes, cause-effect relationships, parent and child processes, and to analyze temporal dependencies.

Using a BIM model, it is easy to see which device requires attention, where it is located, and how it is configured. This integration allows maintenance personnel to easily and intuitively navigate to the device that needs to be maintained, as well as to quickly access all information required for maintenance (see Fig. 9.22).

By accumulating knowledge about the problems that have arisen and their solutions, a kind of technical genetic memory is formed. Patterns are generated from this—a digital model of the "normal behavior" of an object, including a model of the highly efficient energy consumption instead of the usual parameter monitoring in a BMS. If a deviation occurs, the platform automatically sends a message that is added to the CAFM/BPM service request queue.

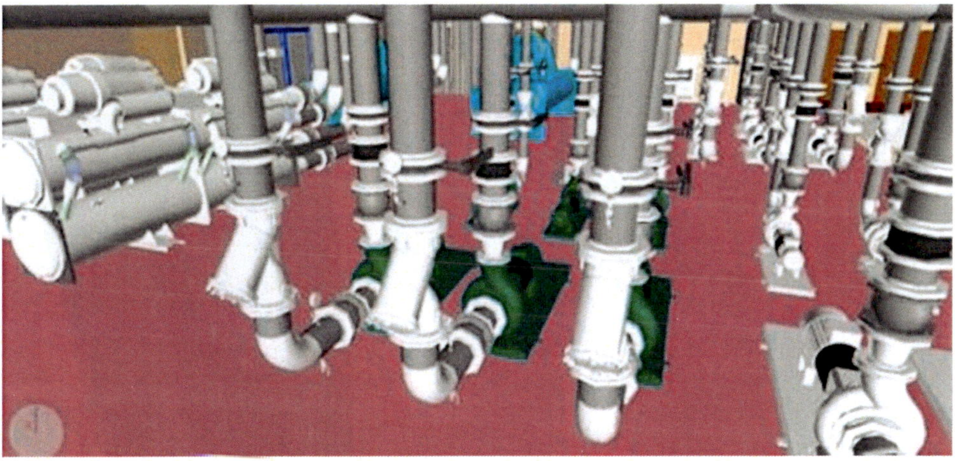

Fig. 9.22 The objects to be observed are highlighted in color in the BIM model

Examples of detecting anomalous asset operation are shown in Fig. 9.23. In the study of the cyclograms of pumps that should normally operate at identical frequencies, anomalous behavior is detected.

One of the pumps ignored the SCADA command to reduce the frequency from 40 to 25 Hz in frequency converter operation. Fig. 9.23 clearly shows the violation of the pattern (pump 7.3, yellow graph). The consumption of the pump has not changed accordingly, even though the SCADA system processed the command. Further analysis showed that a limit of 40 Hz had been set manually on the frequency converter itself. Without big data analysis, this technical error would hardly have been discovered and remedied.

The analysis of the power consumption schedule showed that the commissioning of a compressor led to an uneven operation of the entire system (mismatch of the automation of the compressor control), which is confirmed by frequent short-term shutdowns of the assets that do not last longer than 1–3 min (see Fig. 9.24).

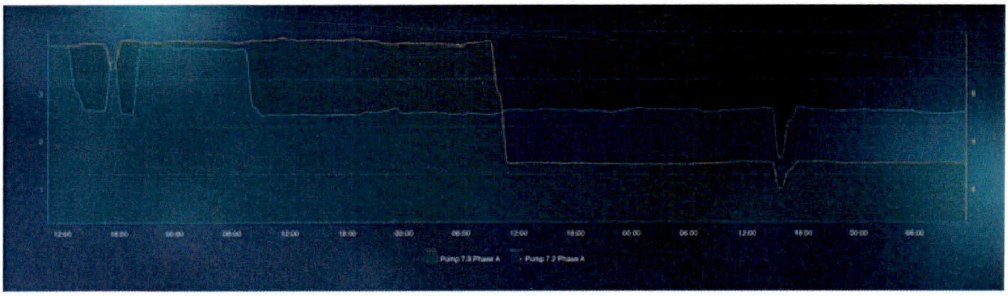

Fig. 9.23 Example 1: SCADA frequency reduction command not executed

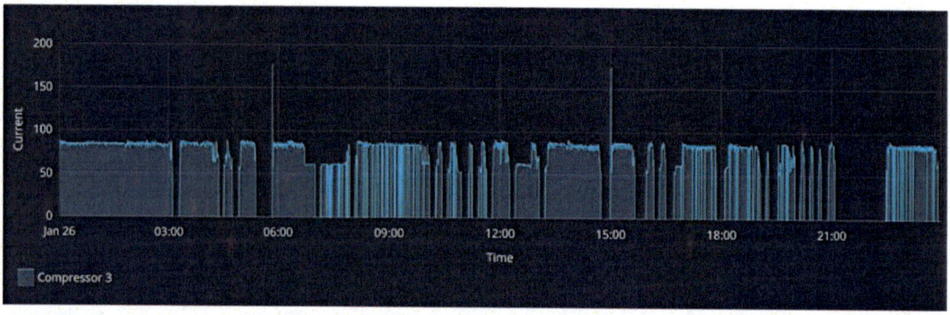

Fig. 9.24 Example 2: One of the three compressors did not work properly

The frequent switching on and off of compressors and the associated start-up currents increase the load on the entire power grid and can lead to premature failure of the expensive compressor devices. A comprehensive analysis was carried out, during which the settings of the automation system were further optimized.

A BIM model can significantly improve the operational efficiency of a building, but time and resources must be spent to supplement the model with technical information on the operation of the assets (which are usually not entered into the model during the planning and construction phase).

In particular, the use of BIM integration in HiPerWare simplifies navigation through the technical equipment and allows maintenance by a smaller number of employees, without the need for them to have extensive experience on site.

A complete BIM model provides immediate access to technical information, maintenance schedules and parameters, as well as operating modes of the assets. The internal development based on IoT technology has simplified the process of capturing and analyzing big data. By combining the BIM model with the HiPerWare platform, a long-term digital footprint of the technical systems was created, which reduces operating costs by up to 30%, improves energy efficiency by up to 20%, and reduces carbon emissions.

9.9 New Construction of an Office Building in the Banking Sector in Prague

9.9.1 The Project

The client of this BIM project, which was about an FM-compliant BIM integration, is one of the three largest banks in the Czech Republic with over 7000 employees. About 100 of them work in the service sector. The operational tasks are carried out by regional or external FM service providers.

The construction of an office building began in 2006, which was not planned with BIM at that time. The personnel scattered in various locations in the Old Town of Prague was to be brought together in one central location. A new headquarters with 80,000 m^2 was created for 2900 employees with LEED Gold certification. But the costs and duration for the acceptance and transfer of the new building into operation ran out of control due to poor or even missing building and asset data. Costs and duration were enormous at that time. Just in the acceptance phase, over 400 working days had to be spent on data preparation and quality assurance.

The number of employees continued to grow—faster than expected. More and more office space had to be rented additionally. Again, a large part of the personnel was distributed over several locations. Therefore, among other things, three decisions were made in 2012: First, a second headquarters should be built. Secondly, one wanted to learn from the mistakes of the past and achieve a better data quality after the planning and construction phase in order to reduce the costs of taking over after completion. And this should be realized thirdly by means of the BIM method and IT in order to transfer the data of the "digital twin" after completion into a CAFM system. In addition, all FM-relevant information that has been collected from the operation of the first headquarters should already be taken into account in the planning.

The objectives were: A new Green Building with 61,000 m^2 area for another 1400 employees and achieving the LEED Platinum certificate. The CO_2 balance should be optimized, new technologies should ensure the lowest possible operating costs, the FM requirements should already be taken into account in the design and construction phase. Finally, the advantages of the BIM methodology should be used for the entire building lifecycle, e.g. with an as-built documentation in the BIM model.

9.9.2 Initial Situation and Approach

Since 2006, the bank has been using the CAFM system Archibus in combination with AutoCAD. Now it was time to extend the existing structure of digital data management systemically and methodologically to BIM already during planning and to implement it for the entire lifecycle. For this purpose, the planning team was extended by a BIM coordinator from Archibus and an FM responsible person from the bank. Corrections, tasks, duties, competences and responsibilities were defined and documented. It was also defined and documented who and for which BIM data is responsible and in which formats and units the relevant data is to be entered. This also included so-called "family sets", classification systems and the expected Level of Development (LoD) in the different phases of modeling.

A main objective was to reduce the considerable effort for the acceptance during the transfer of the new building to operation. A BIM- and FM-compliant planning is important. However, it is not only important to consider the requirements of the future building operation in the initial phase, but also an as-built documentation, which reflects the actual situation at handover. This is then considered a digital twin. The required IT support was enabled by the Archibus-Revit integration.

The requirements for the concept were first specified in the EIR (employer information requirements), on the basis of which the BEP (BIM execution plan) was created.

The concept of BIM2FM included, among other things:

- The unification of terms,
- The agreement on objectives and requirements,
- The benefits of the BIM model in facility management and building operation,
- The requested BIM outputs,
- The requirements for BIM modeling,
- Templates for mapping tables, fields and parameters, technical equipment in Revit and requirements for data synchronization,
- The process and clarification of who delivers which data and parameters at what time in which level of development.

An essential part of the concept was an element-attribute matrix, in which the attributes of all occurring elements were listed line by line and in the columns the time, the necessary LOD and the responsible person were entered.

There was no experience in dealing with BIM models at this point in time. It was decided to not only take the first steps in the planning of the second headquarters, instead the first headquarters was practiced with a BIM model that was created subsequently.

This way, valuable insights were gained into FM-relevant information that should be considered in the planning phase already for later operation.

The data sources for the first model were defined as objects in Archibus and category in Revit:

- DWG drawings stored in the CAFM system,
- As-built DWG's—including small execution changes,
- Alphanumeric data from the Archibus database of the first model with all attributes and parameters per object type.

The same structure was used for the BIM models of both headquarters. This also applies to the equipment and assets as well as to maintenance management with preventive maintenance measures.

9.9.3 Benefits of the BIM-CAFM Integration

The use of the appropriate BIM methodology in combination with a CAFM integration based on a uniform data standard (see Fig. 9.25) was successful. However, the willing-

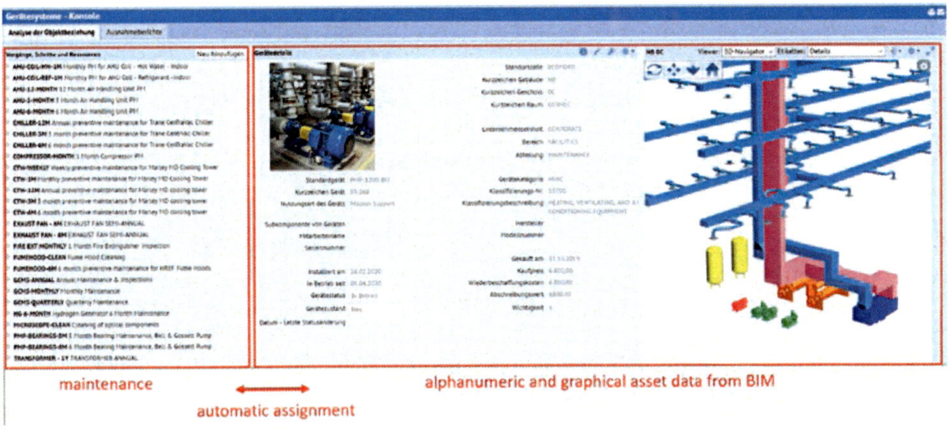

Fig. 9.25 Automatic assignment of data

ness and attitude of the project participants from the different disciplines and responsibilities to collaborate also played a major role.

9.10 Energy Supply and Multi-Service Company in Bologna

9.10.1 The Project

The client is an Italian energy supply and multi-service company based in Bologna, which is active in many Italian municipalities. It provides services in the areas of energy (gas, electricity), water (drinking water supply, including aqueducts), wastewater (sewerage and sewage treatment plants) and environment (waste collection/disposal) for about 4 million citizens.

The participants in the project are listed in Table 9.1.

The project was part of the change management, which included the transformation from a building management based on 2D drawings (CAD) to an integrated facility management with parametric building models realized with BIM.

Within the scope of the project, a test phase was planned based on a pilot building that is realized with BIM (cf. Fig. 9.26) and integrated into the CAFM system Archibus (cf. Fig. 9.27). The focus was mainly on service and maintenance management. The pilot also served as preparation for the future integration of larger projects.

The pilot building represented part of a larger building complex with a gross floor area (GFA) of approx. 2226 m^2, consisting of three levels: basement, ground floor, and attic (accessible). The building is used as an office, with the operating rooms located on the ground floor and the heating system in the attic.

Table 9.1 Stakeholders from the contractor's and client's side

eFM (Contractor)	Client
• Project manager/project contact person • BIM coordinator • BIM expert	• Contract contact person (FM manager) • Project contact person (BIM) • Construction • Maintenance

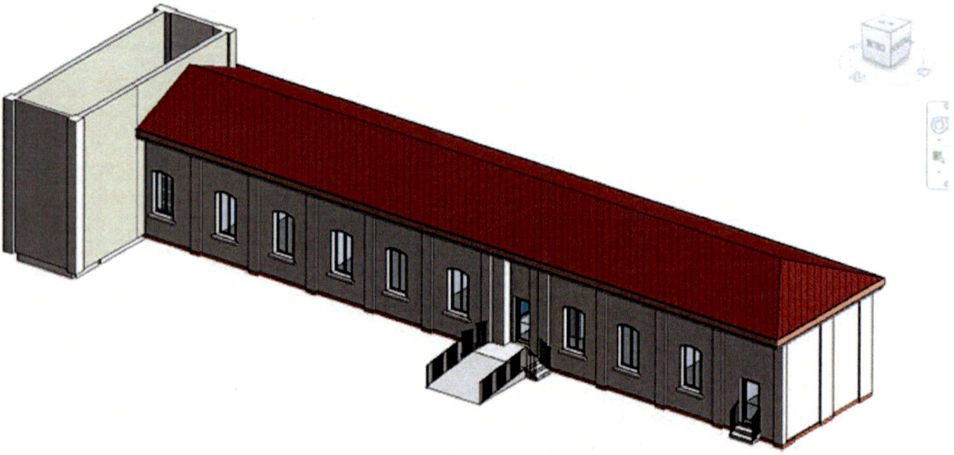

Fig. 9.26 3D view of the architectural model in Autodesk Revit

9.10.2 Initial Situation and Approach

The previous administration of the building and the other facilities were based on CAD drawings imported into the CAFM system Archibus. They were used primarily for space, employee, and move management, but also formed the basis for other FM processes, in particular for preventive and demand-driven maintenance.

The client's requirements were previously specified in the employer information requirements (EIR). They resulted from the desire to further develop in terms of asset management. The aim of the project was therefore the transformation of the methods and tools of operational building management in the area of assets and security. In 2019, the company initiated a BIM planning path, in 2020 BIM was already used on the first construction sites. For 2021, it was planned to start the operation and maintenance activities on the basis of the as-built model available in BIM.

The requirements were therefore specified during implementation of the pilot project in order to conduct a first training on the CAFM system with corresponding FM structure, involving the global service personnel. Specifically, it was asked to select a small building that includes all areas (building, building technology, electrical, fire protection).

The key points of the pilot were as follows:

Fig. 9.27 3D view of the architectural model in Archibus

- Conversion of the building part from CAD to a BIM model and integration into the CAFM system Archibus,
- Training of the FM and global service personnel in BIM management in the field of maintenance for integrating the BIM model into Archibus,
- Design of guidelines for transformation: In Archibus there were already assets managed with graphic elements in CAD or paper form before integrating the BIM model,
- Design of technical specifications to be included in the new global service contract, both for the transformation and for the conversion of assets from Archibus with graphic elements in CAD or paper form into a BIM model.

After the results of the pilot project, coaching was carried out for the FM and global service employees with regard to importing the first buildings modeled in BIM.

The exchange of data related to the managed object properties was limited to the asset components that were the subject of the normal and demand-driven maintenance services.

The BIM-based transformation process of operational building management began with the implementation of a pilot. This resulted in objectives, possible uses and details that are necessary to obtain the BIM models tailored for asset and security management.

The method used to pursue the project goal consisted in dividing the work into four phases, which are summarized in Table 9.2.

Phase 0 was the starting point for the analysis of BIM modeling and the provision of documentation useful for capturing basic information, both geometric (e.g. plans, sections, views) and alphanumeric (e.g. datasheets, maintenance schedules).

Table 9.2 Overview of the project phases

Phase 0	Phase 1	Phase 2	Phase 3
• Prepare documents required for analysis	• Target-actual analysis • Design of guidelines	• Generate BIM model based on CAD floor plans • Integrate BIM model into CAFM system • Train employees	• Optimize results from phase 1 • Create building cluster • Formalize documents to be included in the requirements specification • Support client with integration of the first building managed with BIM into Archibus

The goal of phase 1 was to develop a coherent and sustainable approach to the processes and procedures carried out by the FM department for space and maintenance management, in order to extend the BIM methodology to the technical systems of a building.

The result of this phase was the creation of the specifications for change management—from CAD to BIM. This document represented the guidelines required by EIR for the transformation and was the starting point for the implementation of the asset information model (AIM), applied to the company's buildings, to support asset management.

It included a comparison of the information extracted from the client's documentation with the information originating from the CAFM system and identified to be important for the BIM specifications in asset management.

The as-is analysis focused on two points:

- LOI (Level of Information) analysis—Comparison between the minimum information required by the client in the EIR and the parameters represented and managed in the CAFM system for asset management. The comparison was necessary to identify possible deviations between the parameters and, if necessary, make additions.
- Asset management procedures—Review of the management and maintenance type of each individual asset to identify critical components. The list of elements to be managed in the CAFM system was defined. This did not necessarily match the number of elements to be modeled. In fact, there were asset components present in the BIM model but not subject to maintenance plans and therefore did not need to be integrated. However, their presence in the model was essential for further evaluations such as possible interventions for restructuring, rearranging spaces or moving components.

All the information collected in phase 1 was absolutely necessary to continue with phase 2, which included the creation of the digital model and its integration into the CAFM system.

Before actual modeling, some modeling specifications were defined according to the objective:

- Definition of the modeling object: Definition of the types of building and asset components to be modeled according to the result of the as-is analysis with regard to asset management,
- Definition of the geometric details (Level of Geometry—LOG) according to the requirements of the UNI 11337-4 (UNI—an Italian non-profit standardization organization) for the different types of building and asset components,
- Definition of the alphanumeric details (Level of Information—LOI) according to the requirements of the UNI 11337-6 for the different types of building and asset components, in accordance with what resulted from the LOI analysis (as-is).

The goal of phase 2 was to provide concepts and tools for the generation of a BIM model based on CAD floor plans, the integration of buildings and assets into the CAFM system, and the presentation of the pilot. This resulted in a completed BIM model of the pilot building and the subsequent integration into the CAFM system Archibus.

In order to achieve a correct integration of the Revit model into Archibus, a comparison was necessary that consisted in matching the information present in the database with that of the AutoCAD drawings.

When comparing the data entered in the CAFM system with the object information present in the BIM model, two scenarios emerged:

- Case 1—a perfect match of database information with that of the model,
- Case 2—no match of database information with that of the model due to the following problems:
  - Quantity: More or fewer objects were checked in the program than are actually in stock,
  - Positioning: There was a match of the number of objects recorded in the CAFM database with those in the model, but it was not possible to assign the code of the correct instance because there were several objects of the same type in the same rooms.

The results of the comparison yielded the following:

Total number of elements	74
Number of positioned and Archibus-connected elements	44
Percentage of positioned and Archibus-connected elements	59%

Open issues were resolved through direct consultation with the client. After the model integration was completed, the final phase started.

The aim of phase 3 was to define a BIM implementation plan for expanding the methodology to building assets. This represented the final result of the project. Achieving this goal required several sub-results to be implemented:

- Optimization of the previous results from phase 1,
- Building clusters from which the company's real estate portfolio for conversion into BIM models was composed,
- Formalization of documents to be included in the specifications for the global service,
- Supporting the client in integrating the BIM model into Archibus.

During phase 3, regular coordination meetings were held with the client on the progress of BIM management of the pilot building and the implementation of the guidelines created in phase 1. The reconciled documents were prepared so that they could be integrated into the technical specifications of the new global service contract.

In addition, the drivers for clustering the buildings and creating a plan for converting the remaining building assets were delivered—with the same results as in the pilot case.

Figure 9.28 shows a summary of the conversion process of the building from CAD-based to BIM-based management. The following software programs and functions were used to convert the CAD plans into the BIM model:

- Autodesk Revit—parametric modeling software for generating the digital building model (cf. Fig. 9.29 and 9.30).

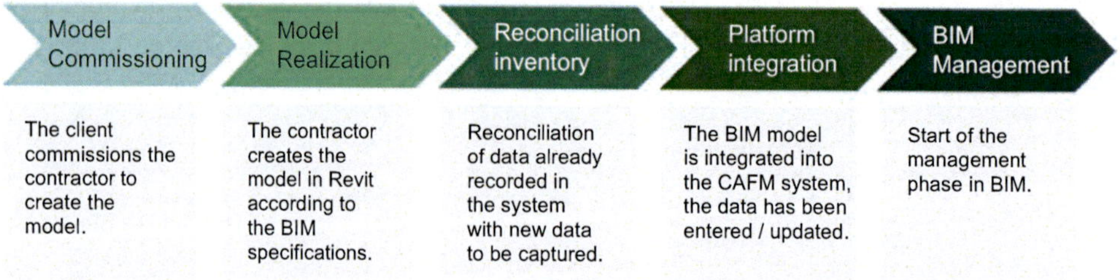

Fig. 9.28 Conversion process

- Archibus—CAFM software as central management platform (cf. Fig. 9.31), with the following main functions for the pilot
  - Configuration of parameters—mapping of attribute fields that are linked to the areas/rooms and assets to be managed in BIM,
  - Assignment—the functionality made it possible to define the type of management of a building and to assign the models to be changed,
  - Upload—enabled the upload of the updated model via the Internet,
  - Reporting—allowed to view the model in 2D and 3D on the Internet, with the possibility of easily exchanging information between different operators and connecting the object data from the BIM model with the FM processes,
  - Archibus PlugIn for Revit—allowed to integrate the model with Archibus and to carry out various operations for updating and synchronizing information and data. The main functions were:

    Assignment of floor plans to the plans available in the database that had to be integrated,

    Data update and information synchronization,

    Validation and publication of the 3D view.

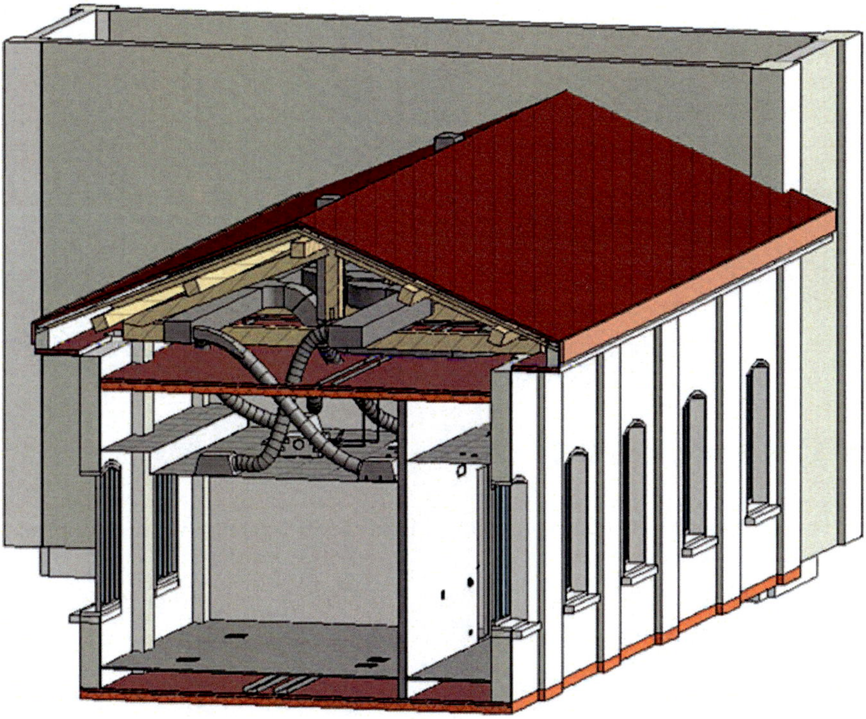

Fig. 9.29 3D section of the coordination model in Revit

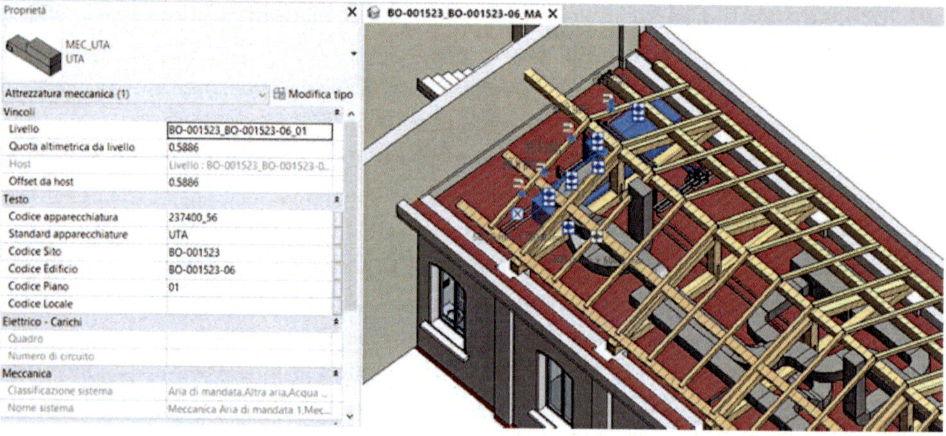

Fig. 9.30 Information detail of a mechanical component (air handling unit) in Revit

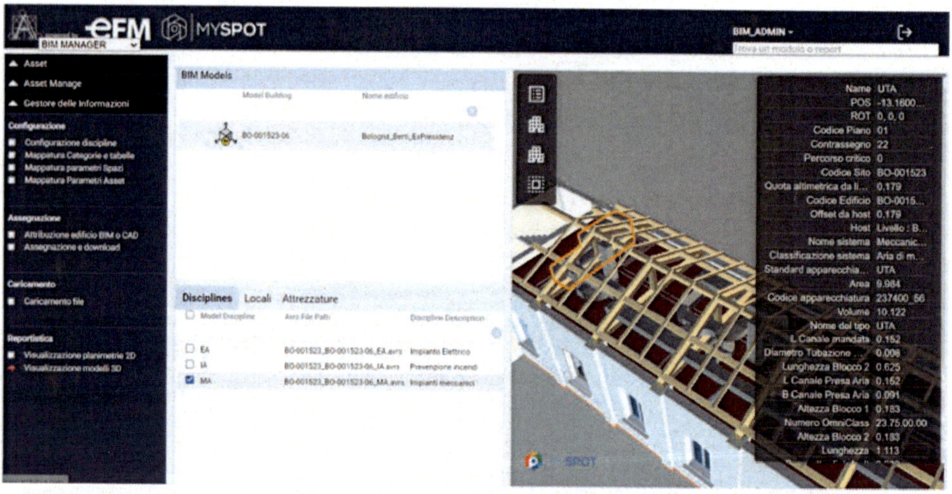

Fig. 9.31 Information detail of a mechanical component (air handling unit) in Archibus

9.10.3 Benefits of the BIM-CAFM Integration

The benefits associated with conversion to a BIM model relate to two main topics:

- Security: The availability of information associated with a virtual model allows for better accessibility in two ways:
  - immediate access to information from a distance,
  - immediate access to information that is not accessible on site,

- Strategic evaluation: By querying a digital model, it is possible to access virtual information that completely describes the product with all its components and properties. This facilitates strategic evaluations that are associated with the relocation of personnel, the rearrangement of spaces or asset components and the restructuring of interventions.

The benefits that can be achieved through the transformation to BIM are numerous and are listed in Table 9.3.

At the end of the transformation, the added value can be summarized as follows:

- Depending on the goal, a certain amount of initial information is required to avoid mismatches when comparing.
- Depending on the goal, it is necessary to determine the degree of geometric and alphanumeric detailing to avoid unnecessary work.
- When managing existing facilities using BIM, the following cases can be distinguished:
  - Building is already modeled in BIM,
  - Building must be reconstructed in BIM,
  - Special case: partial modeling of the building in BIM exists.
- Depending on the type of activity (maintenance, change in legislation, restructuring of spaces), different update processes can be identified in the maintenance phase of BIM models.

9.11 Tempelhof Airport—BIM-based Event Management

9.11.1 The Project

From the perspective of FM, the event management sector poses a special challenge for implementing the BIM method. Thus, it is not only necessary to support the regular operation of a building or an *event location* through BIM models, but also to deal with the numerous short-term changes and adjustments for the planning and execution of individual events. In Sect. 9.11, results from the joint research project "BIM4Event" of the University of Applied Sciences HTW Berlin, the Flughafen Tempelhof Projekt GmbH and the Finnish Metropolia University of Applied Sciences UAS from Helsinki are presented. In this project, agile methods were used for the creation of various BIM models of the former Berlin Tempelhof Airport.

9.11.2 Event Management at Berlin Tempelhof Airport

The complexity of events is constantly increasing due to ever higher requirements in the field of acoustics, lighting or media supply. While it is necessary to deal with these

Table 9.3 Value added by BIM

Area	Benefit
Centralized management of building and asset data	• Detailed knowledge of the nature of surfaces/spaces and assets according to the requirements of facility management • Data transparency for all stakeholders • Immediate availability of data for cost estimation of operational services • Immediate availability of useful information for maintenance personnel • Avoidance of duplicate data entry in master data management of a building
Interoperability between BIM and Archibus models through bidirectional synchronization	• Availability of data in the CAFM system that is synchronized with the BIM models for operational FM • Availability of information and data from the BIM models that can be used by the administration and are accessible to them, which are not directly involved in the creation or editing of models but still in the management of buildings and in capturing spaces and assets • Availability of the latest updated information • Rapidly available information
Comprehensive and in-depth data and information management from the project start	• Complete and timely knowledge of the building and asset stock from the beginning of the project, rapidly available information • Cost savings in terms of retrieving and maintaining asset data
Standardization of management processes	• Optimization of implementing and updating the BIM model thanks to the configuration of the model update and maintenance processes with the definition of the people involved, their tasks and the respective areas of responsibility

mainly technical aspects, higher security and quality requirements must also be taken into account against the background of profitable events. For this purpose, new management approaches are required that go beyond the perspective of classical event management and also include regular FM.

This demanding starting situation requires above all a stringent information management, considering the networked communication relationships of the stakeholders involved in an event and their different interests. Within the framework of the project, two different scenarios were examined. In the first scenario, the in-house FM team is in the focus, which offers rentable spaces and technical support for external event organizers. In the second scenario, the internal FM team is also in the focus, whereby this team is active as part of an inhouse event organization and thus belongs to the event operator.

When examining the first stakeholder scenario, the focus was above all on providing high-quality, billable services for complex events. A special challenge was the flexible handling of initially unclear requirements and their later appropriate implementation in the execution of the event. In order to offer attractive event spaces, the FM team is required to enable technically demanding services reliably and on time. An example of this is the provision of a nationwide and powerful open-air WiFi but also the coordination of short-term, event-related conversions. This includes rapid response for the compilation of building-related information, sometimes as 3D renderings for the marketing of the event location.

In the present research project, the requirements were implemented in the form of short-term "small" construction projects in the building, whereby a nearly complete lifecycle is traversed by the temporary character of the measures (planning, implementation, operation, dismantling). The second scenario considered initially seems to place less demands on the FM and event team, since communication with an external event management is omitted. Here the cooperation takes place mainly internally, whereby simpler approaches to data management were examined. However, by taking over the organizer role, new challenges arise, for example in the commercial field, resulting in interfaces to standard software (e.g. ERP systems) in perspective.

For the two selected scenarios, the former Berlin airport Tempelhof is perfectly suited as a case study, as it has already been used as a venue for external organizers, but also for in-house events (cf. Fig. 9.32).

The figure shows the planned future use of the airport building, which is coordinated with the master plan, with areas for a variety of event types, such as major events, short-term special events in the main hall, but also areas for medium-term and recurring uses in the field of culture, theater and conferences. For the project, it was also advantageous that the Tempelhof Projekt GmbH has its own event team that works with an FM organization partly outsourced and a building department currently occupied with comprehensive renovation measures.

9.11.3 BIM in Event Management

From the perspective of event management, the dynamic adaptation of BIM models for temporary events as well as the complex collaboration during event planning and execution are particularly important. Since there is hardly any research or project experience in the field of BIM for event management so far, the definition of BIM use cases in event management and the resulting information requirements were in the center. In a second

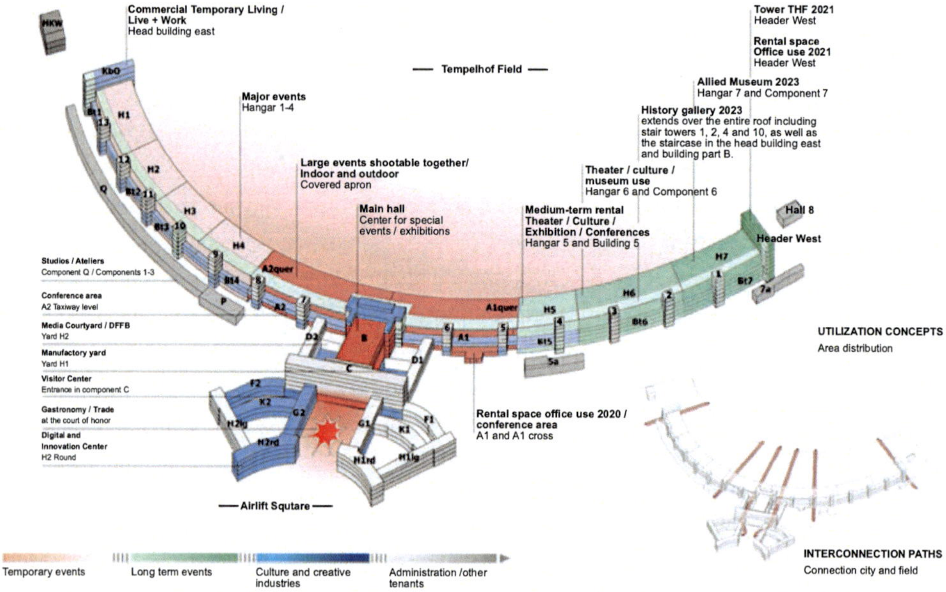

Fig. 9.32 Former Berlin Tempelhof airport with planned uses according to the master plan

step, the workflows of the cooperation of the different partners in the event project were examined.

For long-term use a simplified BIM model of the airport was developed as a *base model* for FM, which was subsequently adapted to the application cases of the two scenarios. Due to the heterogeneous data situation of different parts of the airport building, existing information sources were initially used for the model creation, including 2D floor plans (DWG format) but also a large number of current individual scans of building areas in the form of 3D point clouds. The modeling focused on selected event areas where relevant components (BIM objects) were modeled in more detail. In order to further reduce the effort for model creation, parametric model objects were also combined with 3D point clouds in the BIM model (hybrid models).

An example of a BIM model with fictitious event installations in the main hall (building part B) is shown in Fig. 9.33. The stage installation, aspects of media supply and seating were modeled as an application case. Furthermore, the BIM data were prepared for a crowd simulation to represent evacuation scenarios. For the sake of monument protection, the model and the 3D scan data were combined and provided with realistic textures. In this way, a before-after comparison was possible in order to better assess the restoration of the original state after a temporary event installation. Other BIM use cases were, for example, the model-based determination of quantities and the positioning of event objects.

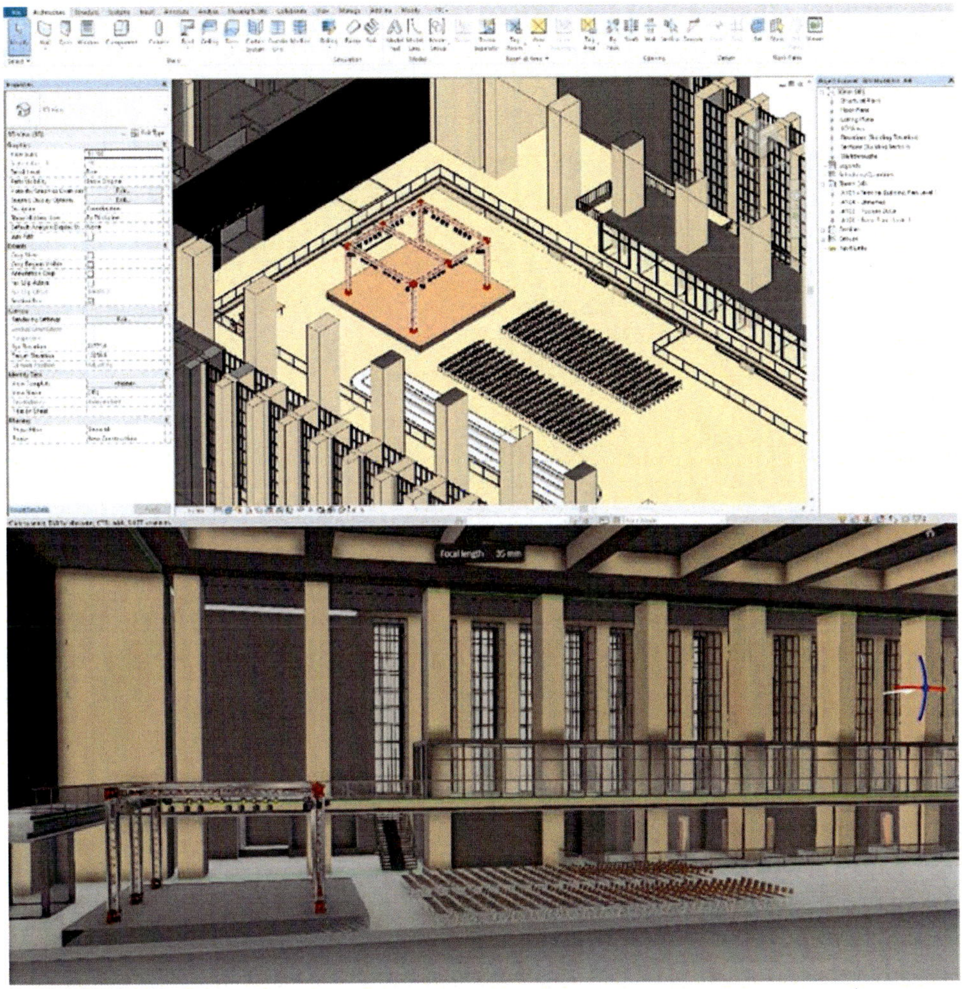

Fig. 9.33 Coordination of event objects in the BIM model with 3D point clouds

9.11.4 Agile Methods for BIM Modeling

As explained in the previous section, different (temporary) event BIM models were derived from an FMBIM base model. These event models provide specific building information locally and support model-based collaboration for the respective application case. However, the definition of the use cases as well as checking the available inventory information and information requirements for model creation only took place during project execution.

At the start of the project, only incomplete requirements were available. Computer Science therefore rely on agile project management methods for such dynamic development projects. It was therefore logical to adapt agile methods from IT for this purpose.

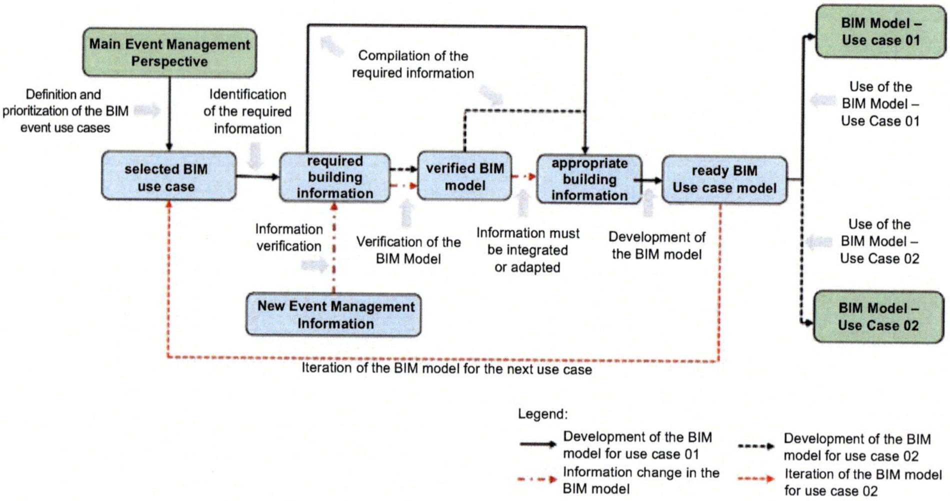

Fig. 9.34 Agile framework for developing different BIM use cases

The virtual character of BIM models as a "product" was beneficial, allowing an iterative and prototypical development comparable to software. Agility from the perspective of project management means that new and changing requirements can be quickly reacted to in a changing environment.

Fig. 9.34 gives an overview of the framework based on the original Scrum approach, which was developed for this purpose in the case study. The starting point is an as-built BIM model, the FM-BIM base model.

In the method specified by the framework, the most important BIM event use case is first determined (e.g. calculation of construction costs or evacuation planning) and the absolutely necessary building information is identified. In the next step, the available information is collected (e.g. number, type, size and position of stage sets and their costs) and, if necessary, a re-capturing is initiated. Subsequently, this information can be integrated into the initial BIM-FM model, resulting in a specific BIM event model. In a validation step, it is checked whether the information mapping is appropriate for the use case, thus completing the first BIM event model and the first sprint. According to further applications (e.g. evacuation planning), the model is now developed iteratively (further sprints). A detailed description of the agile framework can be found in Besenyöi et al. 2018.

9.12 Hochbauamt Graubünden—Administrative Centre "sinergia"

9.12.1 The Project and the BIM2FM Approach

The Hochbauamt Graubünden was able to successfully implement a BIM2FM approach in the construction of the administrative center "sinergia" in Chur (Ashworth and Huber 2021). Fig. 9.35 shows the new building, which was started in March 2017 and occupied in summer 2020.

From the very beginning, the building management was involved in all considerations and decisions. In this project, the facility management (FM) of the Hochbauamt Graubünden was given access to a comprehensive platform for the first time to get involved in project-related and construction-related facility management for the future efficient operation.

Insights into the digitalization in FM played a significant role in the decision to opt for a BIM2FM approach. For example, Schober and Hoff (2016) found in a study that 93% of the players in the FM industry believe that digitalization would influence almost all of their work processes. In addition, it is assumed that BIM has the potential to change the construction industry in the same way that Amazon has revolutionized retail.

Fig. 9.35 The new administrative center "sinergia" in Chur, Switzerland

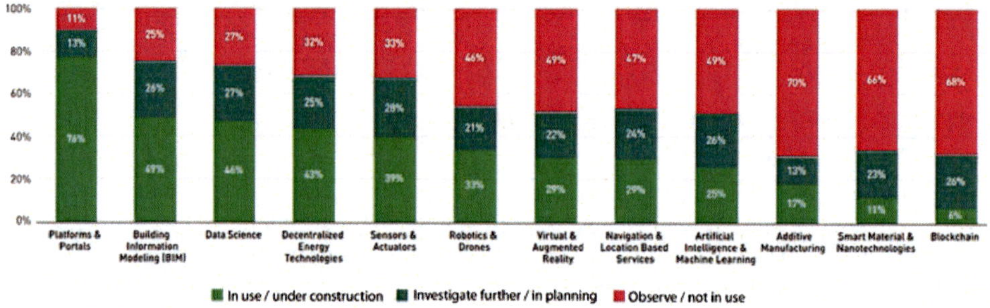

Fig. 9.36 Use of digital technologies in the real estate industry

In a pom+ survey (NN 2020b) *BIM, Platforms & Portals* and *Data Science* were highlighted as the three leading trends that are already being implemented in practice (see Fig. 9.36, NN 2020b).

This is seen as an opportunity to develop new business models, improve existing business processes and make a contribution to the sustainability of buildings. The potential financial savings are also significant. Other studies (Gerbet et al. 2016) predict that full digitalization will lead to annual global cost savings of 13–21% in the design, construction and operation phases by 2025.

9.12.2 BIM Basics in the Project

At the beginning of the project, it was important to create uniform basic principles on the subject of BIM in the Building Construction Office (Hochbauamt). The key to successful BIM projects is that organizations train their teams and take the time to clearly define their BIM goals and the resulting critical information requirements.

It was already emphasized (Ashworth and Heijkoop 2020) how the BIM process can bring considerable benefits to customers and facility managers. It was also emphasized that a BIM project must start with a clear customer vision in order to be successful. This should define what is to be ordered and should formulate realistic expectations. In order to ensure that the result of a BIM project corresponds to what is planned and desired, *information requirements* must be defined as early as possible.

It has been recognized that the use of a consistent terminology in the BIM process is essential. This ensures that all stakeholders develop a common understanding and can therefore find a solution more quickly in the event of problems. The guideline here was the standard ISO 196501:2018 (NN 2018c), with the aim to use the relevant terms and their relationships correctly. This shows the relationship between the types of information requirements (OIR, AIR, PIR and EIR) and the information models (PIM and AIM, see Chap. 1, Sect. 7.2 and Fig. 7.3). By clearly defining the requirements, the construction team has a clear idea of which information and models are to be delivered to the customer.

In reality, the requirements are often defined using *Pain & Gain Workshops* with important stakeholders from the customer's organizations who understand the operational requirements and can help with the definition of OIR, AIR, PIR and EIR. This was also successfully practiced in the case of "sinergia".

9.12.3 The Information Delivery Platform

The Zurich University of Applied Sciences (ZHAW) works together with innovative organizations such as the LIBAL Schweiz AG to provide and use tools such as the new information delivery platform (IDP) IDPPlus. Figure 9.37 shows the information delivery process in this environment.

This also makes it much easier to cooperate with all parties involved in the "sinergia" case. So the whole process is supported digitally in an optimized online format. The IDPPlus platform enables suppliers who may have no or little BIM experience to still meet their contractual requirements. They can provide the required information without the need for special BIM software. Using quality controls in the software, BIM managers can check whether the ordered information is actually delivered. The software also uses Open BIM standards and allows full COBie exports for FM tools such as CAFM.

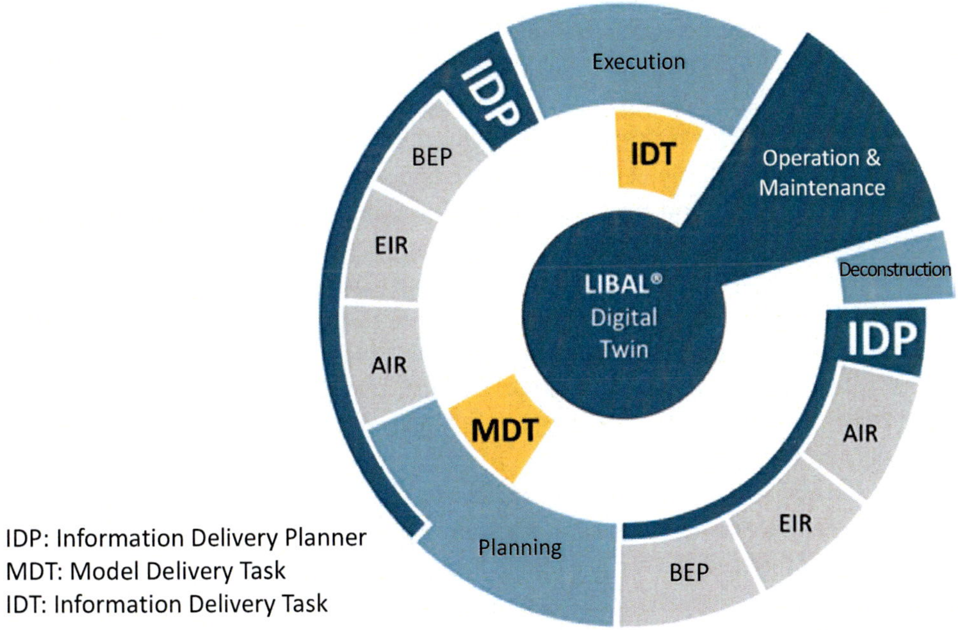

Fig. 9.37 IDP platform for information management in BIM projects

9.12.4 BIM at "sinergia"

In the "sinergia" project, a new approach was developed that includes building management (so-called BIM2FM), which is considered best practice in Switzerland according to the "Bauen digital Schweiz" association.

In the digital building model of "sinergia", all building and building technology components that are regularly maintained or whose condition is monitored due to legal inspection obligations have been classified and provided with relevant information. In total, this is almost 15,000 components. After the data has been transferred to the real estate management system (CAFM system) of the Building Construction Office, these data are linked in an automated process with maintenance schedules, due dates and work estimates. Depending on the type of asset, information on their average lifespan, legal requirements and inspection obligations is also supplemented and is thus available for the operational management of the building.

In this project, the FM of the Hochbauamt Graubünden was given access to the IDP platform for the first time and was able to contribute its ideas to the future operation at an early stage. This early integration of project- and construction-accompanying facility management into the construction planning was very effective. Especially in the early planning phases, the required measures could be incorporated into the project to a large extent from the perspective of operation. Another advantage of this practice is that the largest cost and effort drivers such as cleaning, inspection, maintenance and disposal can be easily identified, so that suitable countermeasures can be taken at an early stage. This helps to reduce lifecycle costs and optimize FM processes.

The information and data management was fundamentally restructured in this project by the Building Construction Office. For this purpose, the documentation model (NN 2013) of the Coordination Conference of the Building and Property Authorities of Public Employers (KBOB) was applied. This model covers the information needs of all those involved in the lifecycle of a property. The complete digital planning with BIM2FM was important for the new orientation of the information and data management and formed the basis for a successful implementation. The operational planning was carried out according to the process model FM of the Building Construction Office (see Fig. 9.38).

The object, cleaning, security and support processes were described and illustrated in the operations management concept. In the past, FM according to EN 15221 focused mainly on "space and infrastructure". With the new construction of the administrative center "sinergia", the main category "people and organization" was also taken into account according to EN 15221.

The process model Facility Management of the Hochbauamt Graubünden was the starting point for the derivation of spaces, tasks and activities. This process-oriented view was the basis for the definition and development of structured data.

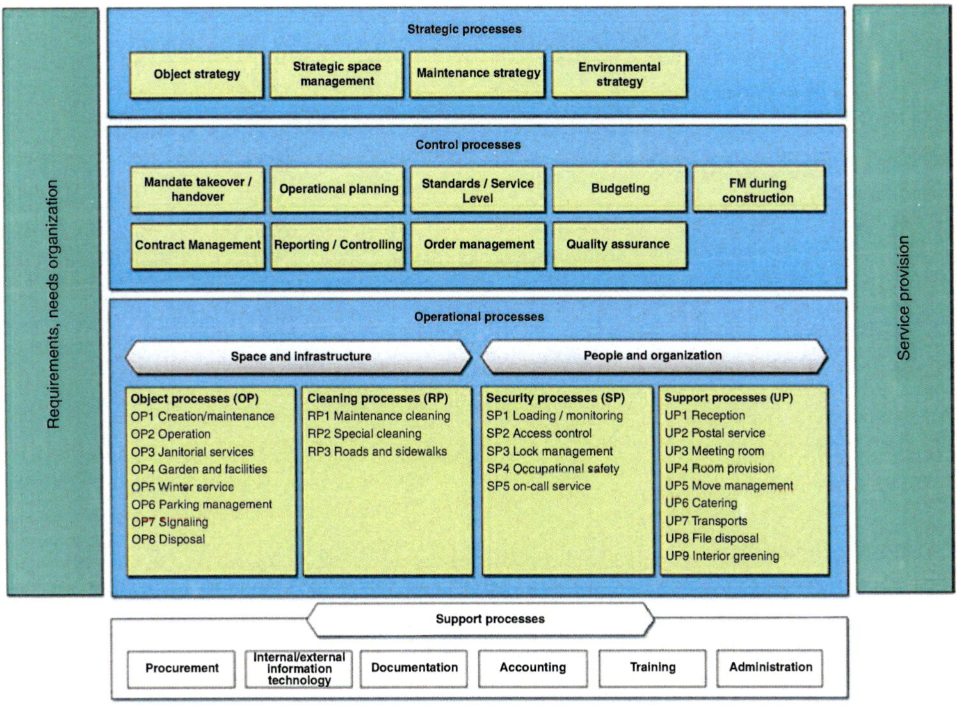

Fig. 9.38 Process model FM of the Hochbauamt Graubünden

9.12.5 BIM-CAFM Integration

To order and provide data in a targeted manner, all FM processes were examined and defined. This analysis was the basis for deciding which data the processes should support in the future. Furthermore, it was decided which data can and should be maintained in the operational phase.

In addition, uniform labeling conventions were necessary. Defined building, floor, room and space names as well as an asset identification system are indispensable and allow for systematic identification and labeling. Corresponding guidelines form the basis for identifying elements in BIM models that are relevant for the operation phase.

Therefore, a CAFM system is used in which all necessary real estate data are collected (cf. Fig. 9.39). With this application, not only all facilities such as objects, rooms, assets and inventory are captured, but also contracts, documents, orders and costs.

The basic data, for example the room structure (spatial hierarchy), rooms with numbering, names, floors, use classification, areas and surfaces, are imported into the CAFM system (cf. Fig. 9.40).

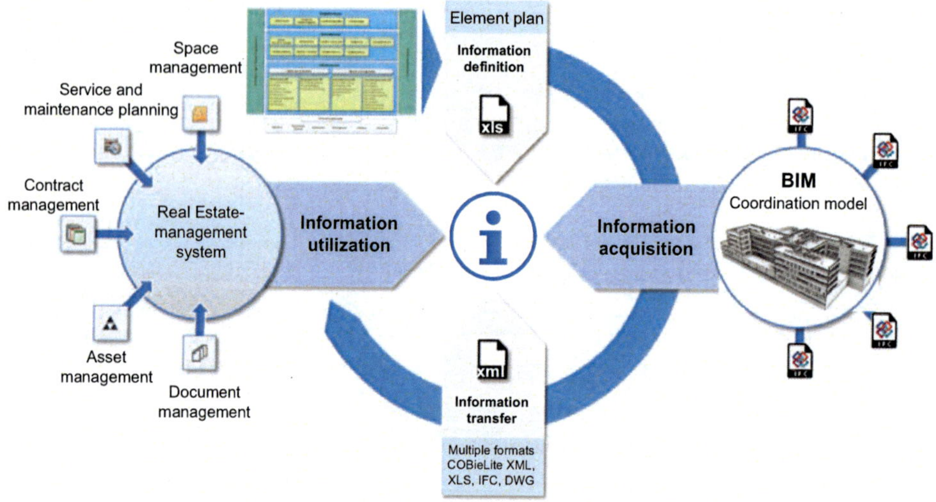

Fig. 9.39 Structure of the real estate management (CAFM) system

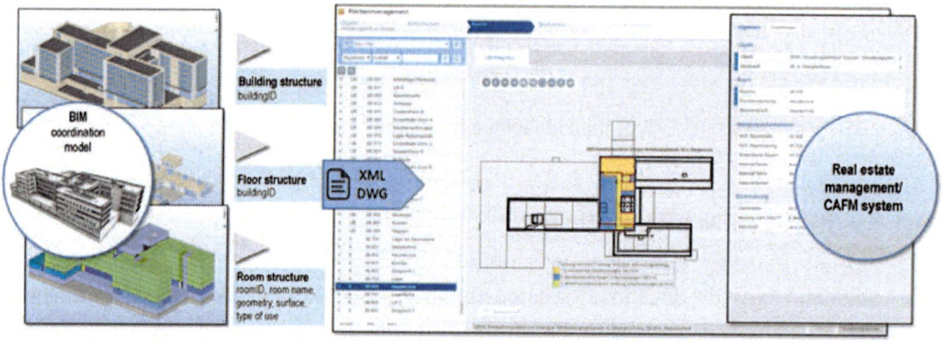

Fig. 9.40 Import of basic data from BIM into the CAFM system

9.12.6 Results and Experiences

On the one hand, spaces form the basis for cleaning and occupancy planning, on the other hand they serve as base data for building technology components. However, the main focus is on the maintenance-relevant parts of MEP technology. This makes it possible to identify all components that, for the reason of owner and operator liability, extended lifecycles or operational safety, require regular attention. The imported components were classified using a catalog. This allows to enrich them with definitions from the maintenance strategy of the Building Construction Office. Furthermore, it is possible to answer questions such as:

- Is a service contract concluded for a certain type of asset or are the maintenance tasks carried out internally?
- Is there a legal obligation to inspect?
- Which maintenance tasks are to be carried out at which intervals?

faster and more precisely than before.

As an important success, it has been shown that BIM2FM enables quick and efficient transfer of data into the CAFM tool in practice. In addition, the advantages of BIM were investigated by using the BIM models for other innovative purposes, including videos for employees for evacuations or medical emergencies.

9.13 Summary

The BIM/FM projects presented in the previous sections have shown that BIM is increasingly finding acceptance in the operational phase—although still quite slowly. Projects with different motivation, complexity, technology, maturity and implementation speed were deliberately selected here. The importance of topics such as use cases, data capture and maintenance, IT integration and BIM know-how was repeatedly emphasized. The case studies are intended to encourage interested parties to take the real estate operation into account right from the start when planning their BIM projects and to consistently use the convincing benefits of coupling BIM and CAFM tools.

The projects considered are summarized below.

Municipal Real Estate Jena
The project shows, using the municipal-owned KIJ of the city of Jena, how planning data from Revit is transferred into the CAFM system SPARTACUS Facility Management and how the BIM model is used during the operational phase.

Axel Springer New Building in Berlin
The goal was to create a continuous building model that from the outset had a focus on BIM and its added value during the operational phase. By integrating with ERP and BMS, a digital twin of the building could be generated. The CAFM system used is pit—FM.

Museum of Natural History Berlin
The goal is to resolve the existing backlog of renovation work, to repair damage from World War II and to modernize the museum. The research project of the museum and HTW Berlin supports the transfer of BIM data from planning to a CAFM system as an as-built model and thus to gain early insights for FM.

ProSiebenSat.1—Mediapark Unterföhring

The project of the ProSiebenSat.1 Group is presented, which plans the construction of a new campus in Unterföhring based on the BIM methodology. The aim is to create continuous data flows throughout the lifecycle. Data are transferred for the operational phase to the CAFM software eTask.

BASF in Ludwigshafen

For the new construction of a laboratory building at BASF, a holistic building model was created using the BIM methodology with the goal of using BIM for planning, implementation and transfer of data into the use phase. Closed BIM and Revit were used.

TÜV SÜD in Singapore

The TÜV SÜD location in Singapore moved into a new integrated laboratory and office building in 2021. The BIM methodology was used intensively during construction. The regular comparison of 3D scans with the BIM model led to an optimized construction phase. The transfer of data into the operational phase could be successfully implemented using BIM2CAFM data standards.

Country Park III in Moscow

This is a development project of the BPS Group, which comprises a 22-story office tower. The goal of the project was the use of a BIM model throughout the lifecycle of the building. Revit was used for planning. The digital twin was implemented with the HiPerWare platform developed by BPS Group.

New construction of an office building in the banking sector in Prague

This is a new construction project of an office building in Prague with early integration of facility management. The implementation is carried out with Archibus-Revit interface and a BIM2FM concept.

Energy supply and multi-service company in Bologna

This case study presents a complex transformation project with the goal of converting traditional CAD documentation to BIM. Based on Revit and Archibus, the transformation project including the change process was started.

Tempelhof Airport—BIM-based event management

In a research project, the BIM method was used for event management and was evaluated subsequently.

Hochbauamt Graubünden—administrative building "sinergia"

In the new construction of the administrative center "sinergia" in Chur, the Building Construction Office Graubünden was able to successfully implement a BIM2FM approach. By early integration of facility management, the data could be transferred directly to the CAFM software via a BIM2FM interface.

References

Ashworth S, Heijkoop A (2020) Bestellerkompetenz: Kritische Erfolgsfaktoren für ein BIM-Projekt. fmpro service (2020)1, 32–35

Ashworth S, Huber M (2021) BIM2FM. fmpro service (2021)1, 21–23

Besenyöi Z, Krämer, M, Faraz F (2018) Building Information Modelling in Agile Environments – an Example of Event Management at the Airport of Tempelhof. IPICSE-2018, MATEC Web of Conf. Vol. 251, 2018, 10 S

Gerbet P, Castagnino S, Rothballer C, Renz A, Filitz R (2016) Digital in Engineering and Construction: The Transformative Power of Building Information Modeling. http://futureofconstruction.org/content/uploads/2016/09/BCG-Digital-in-Engineering-and-Construction-Mar-2016.pdf (retrieved: 08.09.2021)

NN (2013) Bauwerksdokumentation im Hochbau – Dokumentationsmodell BDM13, IPB – KBOB, 28 S

NN (2014b) PAS 1192-2 (2014) Specification for information management for the capital/delivery phase of construction projects using building information modelling. London: British Standards Institution

NN (2018c) ISO 19650-1:2018: Organization and digitization of information about buildings and civil engineering works, including building information modelling (BIM) – Information management using building information modelling: Part 1: Concepts and principles. https://www.iso.org/standard/68078.html (retrieved: 14.10.2021)

NN (2020b) Digitalisierung der Immobilienwirtschaft: Digitale Immobilienverwaltung: 2020 Schweiz und Deutschland. pom+ Report. https://www.digitalrealestate.ch/products/digital-real-estate-index-2020/ (retrieved: 08.09.2021)

NN (2021s) Zukunftsplan des Museum für Naturkunde Berlin. https://www.museumfuernaturkunde.berlin/sites/default/files/mfn_zukunftsplan_digital.pdf (retrieved: 17.09.2021)

NN (2021ab) Kommunale Immobilien Jena. http://www.kij.de (retrieved: 26.09.2021)

NN (2021ao) Unternehmensvorstellung. https://www.prosiebensat1.com/ueber-prosiebensat-1/wer-wir-sind/unternehmensportraet (retrieved: 01.11.2021)

Schober K-S, Hoff P (2016) Think Act – Beyond Mainstream: Digitalisierung in der Bauindustrie – Der europäische Weg zu „Construction 4.0“. Roland Berger GmbH. https://www.rolandberger.com/publications/publication_pdf/roland_berger_digitalisierung_bauwirtschaft_final.pdf (retrieved: 20.08.2021)

10 BIM Perspectives in Real Estate Operations

Markus Krämer, Simon Ashworth, Michael Härtig, Michael May and Maik Schlundt

10.1 Critical View of BIM

BIM is critical to digital transformation in the construction industry comparable to Industry 4.0 in manufacturing. Digitalization of planning processes in construction projects leads to optimized business processes. The BIM method is already successfully used in the design, planning and construction phases of a building. This is due to the fact that the increasing complexity of planning and construction projects is very challenging using traditional methods.

The situation is different in operation; due to various obstacles, the application of BIM is not yet widespread in the operational phase. In many cases BIM has presented

M. Krämer (✉)
Hochschule für Technik und Wirtschaft Berlin, Berlin, Germany
e-mail: markus.kraemer@htw-berlin.de

S. Ashworth
Zürcher Hochschule für Angewandte Wissenschaften (ZHAW), Wädenswil, Switzerland
e-mail: ashw@zhaw.ch

M. Härtig
N+P Informationssysteme GmbH, Meerane, Germany
e-mail: michael.haertig@nupis.de

M. May
Deutscher Verband für Facility Management (GEFMA), Bonn, Germany
e-mail: michael.may@gefma.de

M. Schlundt
DKB Service GmbH, Berlin, Germany
e-mail: maik.schlundt@dkb-service.de

M. May et al. (eds.), *BIM in Real Estate Operations*,
https://doi.org/10.1007/978-3-658-40830-5_10

a challenge as operational teams are naturally behind their construction colleagues in acquiring the right BIM competences. Another issue is the lack of standardized data models for objects used in building operations. Nor are there reliable, standardized interfaces for communication between the software products involved. The large number of individual products and pragmatic solutions based on tools such as MS-Excel in operations exacerbates the problem even further.

It is noticeable that in recent years available CAFM systems have integrated BIM features such as IFC import (cf. Sect. 4.3.2.1) and interfaces to various BIM authoring tools (cf. Sect. 4.3.2.2). However, when questioned most CAFM vendors can usually only refer to a few BIM-CAFM projects (cf. Chap. 9) that have actually been implemented so far. It must be noted that there is currently high interest in BIM from clients and building owners in real estate (RE) management, but the willingness to invest is not always on the same level. According to such stakeholders, BIM integrations are often too complicated, not enough standardized and have unclear or non-existent business cases. In addition, there is the hurdle that the application of BIM is usually only economically feasible for new buildings. The many existing buildings often do not justify subsequent BIM modeling, as today's automated digital capturing methods still require considerable modeling skills requiring manual work (cf. Sect. 5.2). Frequent problems with current software products, such as faulty, unstable software versions, insufficient usability or performance problems when loading larger BIM models, strengthen this perception.

There are currently only few trained BIM experts, especially in the field of FM. The development and introduction of uniform BIM data standards across the entire building lifecycle is far from complete, even though there are currently some remarkable initiatives (see Sect. 8.4). As with the application of any new method, the economic viability must be checked and proven in each individual case. The actual use of BIM in a project can be very different from the facility manager's point of view (see Chap. 7), where an over-comprehensive application of BIM can sometimes be counterproductive. In particular, if very large, complex data sets are generated by the BIM applications, they often cannot be maintained in a comprehensive way over the period of operations, so that "data garbage" quickly accumulates.

The greatest benefits and synergy effects of BIM always arise when BIM models are really passed on from planning and construction to the operational phase of FM. So, the transfer of spaces (e.g. floor spaces or window surfaces) for space calculations (e.g. of cleaning areas) is already a widely accepted BIM use case. But even in this case, numerous detailed information from the BIM models are of no value for the operational phase and must be filtered before the transfer to FM. This often causes considerable effort, especially with complex models. Only if information requirements of FM are fixed from the beginning of a project, can corresponding attributes and characteristics be specified and the desired benefits be accomplished. The same also applies to the takeover of the technical building documentation. Here, standards for data exchange must be established and defined in the project.

Further advantages arise from an improved FM-compliant planning, since all disciplinary planners can work together using coordination models and thus enable early testings from the perspective of RE management and FM. These can be collision detection, checks of workspace for maintenance activities or cost simulations of the operational phase. All this together then results in the desired high economic benefit of BIM (cf. Sect. 3.3 and Chap. 6). As an example in Germany, the focus is often still on the construction phase of the building and only rarely on operations. However, an FM-compliant planning offers considerable added value for operations. So planning errors that would only be discovered during operations can be avoided from the outset. This however varies from country to country.

As a consequence, it is important to plan providing a consistent data basis for FM often referred to as a single source of truth by BIM. This allows FM data to be checked and made available much earlier for use in the FM processes. In this way a smooth flow within operations can be achieved and positively affect all subsequent FM business processes such as rental, key management, space management and maintenance.

In this way, tenders for facility services can be carried out much faster and more accurate on the basis of reliable data by using BIM model data. Also in this example, the key to success is the early involvement of FM in the overall process, the requirements of which must be known and contractually agreed by planners and contractors as early as in the planning phases. FM employees must receive corresponding further training in the field of BIM application areas for this purpose.

Another challenge is the continuous update of BIM data, because changes to the BIM models arise inevitably from operational processes. Whether and how these can be sensibly incorporated in practice is not yet completely clarified. In addition, since facility managers are usually not experts in the use of BIM authoring tools and often building owners also have no corresponding know-how this further complicates the matter. As a result, in practice there is often a discrepancy between the status of the BIM models and reality. If this is not proactively managed, the BIM data can no longer be trusted and ultimately two separate data worlds exist again, as was often the case in the past.

Overall, the development of BIM to address the above discussed issues remains exciting and continues to offer interesting opportunities.

10.2 Research on BIM in Real Estate Operations

In the previous sections, the application of the BIM method in FM was in focus and thus the state of the art. The significant interest in BIM in research and practice has led to numerous activities in research and standardization such as norms and guidelines, which has already delivered first results already implemented in practice. It should be noted, however, that most universities and research institutions still focus on the phases of planning and construction.

This section sets out to provide an overview of current and future research topics in the context of BIM in RE and FM. In the following, important fields of research will be briefly explained and exemplary initiatives and research projects will be presented.

10.2.1 BIM Standardization

The starting point for many research and standardization activities was ensuring a smooth data exchange between a BIM-based planning and construction process and FM. In this context, numerous initiatives have been carried out by associations such as GEFMA, RealFM, VDI and buildingSMART, but also by the standardization institutes ISO, CEN, BSI and the German DIN. A good overview of these initiatives and the resulting standards can be found in Sect. 5.1 and Appendix 2 as well as in Bartels (2020).

The problem of using different classification systems to describe and identify building components and their different data formats continues to complicate data exchange also with FM. With the initiative described in Sect. 8.4 *IBPDI* (International Building Performance & Data Initiative, NN 2021t) a universal data model for the real estate industry (Common Data Model for Real Estate) is being created. Another initiative is *BIMeta* (NN 2021az), which is developing an open, vendor-independent digital platform to link BIM class definitions and their properties according to DIN EN ISO 23386 (NN 2020e) from different standards and guidelines in a uniform system.

While research activities in the field of FM handover focused on the end of the construction phase in the past, research activities are increasingly shifting to collaborative data exchange during the planning and construction phases with FM. An example of this is Bartels (2020), who describes a data exchange structure model in FM that has been extended with dynamic, process-oriented data based on IFC. As part of the research project BIM-based operations (NN 2019l)—funded by the BBSR as part of the German initiative *Zukunft Bau*—in addition to detailed future BIM process chains, numerous new BIM profiles were also created for exchanging BIM data with CAFM systems via the CAFM-Connect interface (see Sect. 5.2). Research in this area is being pursued in numerous projects.

10.2.2 Digital Capturing of Existing Buildings

One of the major challenges in using BIM in the operational phase results from the fact that for more than 90% of the existing buildings to be managed, no digital building models are available yet and the existing documentation is also often incomplete. This gives rise to the research question of how to simplify and automate the capturing of digital building models of existing buildings.

In recent years, therefore, a number of research initiatives have dealt with digital capturing methods, in particular 3D laser scanning and photogrammetric methods for existing buildings. Initially, the focus of research was on the workflow for transforming scan results (3D point clouds) of the capturing technologies into parametric, digital building models according to the principle of Scan2BIM or the further use of 3D point clouds linked to parametric digital building models, so-called hybrid models (Krämer and Besenyöi 2018).

Another important area of research focuses on the automatic evaluation of the recorded images or 3D point clouds. For example, researchers at the University of Weimar (Hallermann et al. 2019) are working on the automatic determination of the condition of buildings using unmanned aircraft systems (UAS). Increasingly, methods of artificial intelligence (AI) are also being used to automatically transform existing 2D drawings into digital building models and to automatically recognize BIM objects such as walls, piping systems or MEP equipments in 3D point clouds. Reference should also be made to the research project BIMKIT, which started in 2021, in which, based on the European cloud initiative GAIA-X, the modeling of existing buildings and infrastructure using AI is being investigated in order to generate digital twins. Automated 3D building modeling was also investigated in the DFG-funded project MAV4BIM, in which camera-supported micro-flight robots with photogrammetric methods were utilized.

10.2.3 Common Data Environment, Linked Data and Digital Twin

Common Data Environments (CDE) were intensively discussed in Sec. 4.3. With the switch to the operational phase of buildings, the focus shifts from project-related to asset-oriented data management and thus also the focus of a CDE. Meanwhile, commercial CDE products are also available for the operational phase, often in connection with CAFM systems. The *IBPDI* initiative was already referred to in Sect. 10.2.1 for a new approach to cross-company collaborative platforms.

A research focus in the field of CDE from the perspective of FM is the linked data approach for CDE's. With this approach, the integration system does not store any data itself, but is able to answer queries by a virtual integration. Rather, the integration system has enough metadata to find and query the associated data sources at runtime. This approach is particularly suitable for the integration of very heterogeneous data sources, such as those created by federated BIM models of different disciplines, digital 3D point clouds from 3D terrestrial laser scans and operational IT systems, such as CAFM, CMM or ERP systems.

The project *BIM-FM* supported by the Berlin IFAF institute is given as an example of research initiatives in this field. In this project, two digital building models for existing buildings (St. Hedwig Hospital Berlin, Verbändehaus Berlin) were generated with the support of digital capture technologies. Furthermore, data management was investigated using the linked data approach (Krämer et al. 2018). For the resulting prototype CDE from the operator's perspective, various (multiperspective) IFC models were linked virtually with commercial CAFM systems including a basic management of point cloud segments. Different database technologies were used for this. For example, the integrated CAFM systems were based on common relational database management systems (e.g. Microsoft SQL Server). The as-built models created with the help of the Scan2BIM workflow were stored in the IFC format with the help of the OpenBIM server (formerly IFC server) based on the key-value store technology. The scan results of the digital capture technologies (3D laser scanner, surveying drone, smartphone) were again mapped as point cloud segments enriched with markups in the E57 and LAS format. The actual virtual linking was carried out with the help of the semantic web technology, namely the Resource Description Framework (RDF) using the NoSQL triplestore Fuseki and the ONTOP framework as a SPARQL endpoint for the SQL databases.

With this CDE setup based on the linked data approach, the three information sources CAFM database, IFC models and 3D point cloud segments could be loosely interlinked and yet queried together. For example, a query for a certain fire extinguisher returns information about the maintenance status of the fire extinguisher from the CAFM system, the location in the building from the BIM model or, if no fire extinguishers are modeled in the BIM model, markups of the point cloud segment of as-built scans. This could be used as a demonstration that virtual linking of CAFM systems with BIM models is also possible without direct synchronization of BIM objects with entities within the CAFM database that is still common today.

Further research in this area uses virtual integration based on linked data to create digital twins. In the project *BIM2TWIN* (duration until 2024, NN 2021x), which is funded by the European Union under the Horizon 2020 program, a digital building twin platform is being created. Here, machine learning techniques are used to achieve efficiency gains through the avoidance of waste and thus cost savings, quality improvements and a reduction of the carbon footprint. The research project *BIM2digital-TWIN* (NN 2021y) has also dealt with the creation of digital twins, in which the German Council of Shopping Places (GCSP) and the Bergische Universität Wuppertal have been researching real estate asset management processes in commercial real estate.

The project *BIMSWARM* (NN 2021z) also has the goal of developing an open platform for construction projects, with a focus on tool chains of certified applications, services and catalogs in order to enable seamless and project-specific digital value chains. The research focus on CDE and linked data is indirectly addressed by the project, as the flexible collaboration of those involved in the construction process is in focus.

10.2.4 Visualization, Virtual and Augmented Reality

Two research activities are to be presented as examples, in which visualization and augmented reality (AR) (cf. also Sect. 2.6) play a major role in combination with BIM.

The progress observed for some time now in maintenance management and other related tasks in FM is characterized by a high demand for information. Many potential benefits often can not be achieved today because there are still major problems in providing the information required on site. Only by developing end devices for the location-independent, clear provision of this information and the associated software can maintenance management and similar tasks be carried out faster, more safely and more efficiently. This was the starting point for the research project *FMstar* (cf. May 2017; May et al. 2017), which stands for "Facility Management with the help of semantic technologies and augmented reality". *FMstar* aimed to connect the virtual and the real world in an innovative way using modern AR technology and make it usable for complex FM processes during approval and maintenance. The core of the project was to enrich technical environments in manufacturing by AR on mobile devices such as tablets or smartphones with FM data. Intuitively understandable information such as visualizations and 3D models as well as manuals and technical data help to make context-based decisions in complex FM processes.

The required asset data and the 3D model are extracted from BIM models in IFC format and partly from CAFM models. The respective data sources are imported according to the corresponding modeling standards with the help of plugins, with the graphical model data being stored in a 3D database and the FM master data and process data being mapped in a semantic database. The mobile app developed makes it possible to retrieve relevant context-based planning and status data of technical assets on site on mobile devices. These are then mapped onto the mobile device as an overlay to the images of the real asset and equipment captured by the camera and thus form a direct link between digital information and the real world. The assistance function provides navigation, work instructions and documents, while the user can move freely in the real space. 3D models are moved according to their own position and enable the specific selection of certain components to retrieve their data. For example, the user identifies objects or technical assets by means of QR codes and the scanning of documents, images, etc. The semantic database identifies the information required, which can be displayed and also edited on site.

Innovative IT-based teaching and learning concepts in FM are still in their infancy, with the exception of e-learning platforms used in universities and isolated business simulation games. In the field of further education, these approaches are still largely unknown.

This was the motivation for the project "PlayFM—Serious games for IT-supported knowledge transfer in facility management" (May 2013; Salzmann and May 2016) funded by the German Federal Ministry of Education and Research (BMBF). The meaningful use of serious games within the framework of game based learning (GBL) opens up new potentials and exciting challenges in knowledge transfer. The aim of the PlayFM project was to develop GBL concepts and methods for knowledge transfer in FM in a holistic way and to implement them prototypically in a computer-supported serious game playFM. This was one of the first approaches to apply GBL in a highly complex area such as FM. Target groups are FM service providers, FM specialists and students as well as the management level in companies. The software architecture of playFM is divided into the actual game program, which is installed on a client computer, and the game server, which contains the database with game states and configuration files and the web server with the web-based configuration interface and a high score table. The implementation concept provides for a flexible integration of the playFM learning content, which can be inserted into the game via a web interface. After an extensive analysis, the decision was made for the game development environment (game engine) Unity3D, which allows for fast prototyping. In addition to the classic functions such as a physics engine, sound support, particle effects or 3D import and export functions, the engine has all the features required for the game. The games created with Unity can be run on different platforms such as PC and mobile operating systems, game consoles and popular web browsers. Here also is a relationship to BIM. The further development "Unity Reflect" today includes a suite of products that allow BIM data, stakeholders and every phase of the lifecycle in architecture, engineering and construction to be connected in an immersive real-time platform. It provides a solution for design and coordination that connects all project participants on an immersive, collaborative real-time platform, regardless of device, model size or geographic location. It is also used as a plugin in commercial BIM authoring tools.

10.2.5 FM Knowlege Management and Artifical Intelligence

Making experiences and knowledge about the operation of buildings by FM organizations and FM service companies systematically available for decisions and thus for future action of planners and operators is and remains an important field of research. The use of BIM in the context of knowledge management (KM) influences FM, not only by possibilities for localization and visualization, but also by comprehensive semantic content and relationships that digital building models represent.

The objective of the BMBF research project "*FM-ASSIST*—Computer-aided assistance system for complex decision-making processes in FM" (May and Bernhold 2009) was to support consulting activities with the help of an IT-based assistance system. Can the tasks that are usually taken over by consultants be transferred to a knowledge-based IT system at all? Or exaggerated: Will consultants still be needed in the future or can they be better supported when serving consulting their customers? The project idea was driven by the conviction that consultants will certainly not become superfluous, but that nevertheless important tasks can be transferred to a suitable IT system. Also, the early use of an assistance system can help to approach projects in such a way that more errors are avoided or reduced and consulting costs are reduced. The scenario chosen was the knowledge-based support of companies and public administrations when implementing FM in their organizations. For this purpose, a general procedure model for the introduction of FM was developed and described in the GEFMA guideline 110 (NN 2009a). This introduction model comprises more than 70 sub-processes with corresponding extensive recommendations for action and tools. Because of the desired separation of data, functional logic and user interface as well as the required access via an Internet browser, a multi-layer software architecture was chosen. Two different user interfaces were provided. The expert interface serves the FM expert to enter knowledge as FM procedure model and other knowledge components into the system in a structured way. The FM user interface represents the actual assistance system. Here, an interactive questionnaire is carried out on the basis of the respective procedure model. In addition, as a result the corresponding recommendations for action are generated and further information is provided. The underlying method of the implemenation is similar to case-based reasoning (CBR). Today, with the enormous progress in AI, new methods such as machine learning are available with which specific FM implementation projects can be analyzed and used for predictions.

Similarly, Motawa and Almarshad (2013) have also developed a knowledge-based IT system that helps in acquiring and retrieving knowledge about maintenance activities of buildings. However, here the developed CBR module was connected with a BIM, so that the information about maintained components, which were defined by the CBR module, could be retrieved and visualized by the BIM module. Consequently, this research has raised the question of how the principles of knowledge management, which are embedded in knowledge-based IT systems, can be connected with the principles of information management, which are available in BIM systems for FM.

The great potential of linking KM and BIM is shown by the numerous developments of BIM checking tools (cf. Sect. 4.2), such as *Solibri*, which increasingly offer rule-based checks in the field of FM, e.g. for compliance with floor space standards or distance rules for escape and rescue routes. But this is certainly only the beginning of the possible uses of a combination of KM and BIM.

Accordingly, Besenyöi and Krämer (2021) have examined how, by means of an ontology-based approach, definitions of the most important principles of BIM-based knowledge management and BIM-based knowledge management systems for FM can be gained. As a result, an ontology-based framework emerged, which both defines the necessary tasks of knowledge management and maps the various BIM-supported technologies and tools that support the respective task. The framework shows which methods of knowledge-based IT systems should be developed in the context of BIM in order to fulfill these predetermined tasks today or in the future.

10.2.6 Sustainability, Energy Efficiency and CO_2 Optimization

Challenging goals agreed in the Paris Climate Accords in order to counteract global climate change, have helped drive research in the field of sustainability and energy efficiency which has gained increasing importance internationally. In order to limit the increase in global warming to 1.5 °C, extraordinary efforts and progress must be made in the field of sustainability and energy efficiency in the building sector, especially for existing buildings. However, this will only be possible if facility managers and thus building operations are systematically involved in measures to improve sustainability and energy efficiency.

There are numerous research initiatives to improve sustainability and energy efficiency in the building sector and they cover almost all disciplines of knowledge. Thus, the research priorities already presented, e.g. in the area of FM knowledge management (KM) or digital twins, contribute to the optimization of building operations and thus to the reduction of energy consumption by specifically providing experience and knowledge of FM organizations from building operations to planning and operations. A comprehensive overview of all approaches to sustainability cannot be given within the scope of this section. Rather, some research activities are highlighted in which the BIM method contributes to the achievement of sustainability goals.

One of the central points of focus for the contribution of BIM in the field of sustainability is the use of digital building models as a systematic information and knowledge repository. The project *Life Cycle Assessment and BIM in Sustainable Building* funded by the BMI (Lambertz et al. 2019) aimes at simplifying the ecological assessment of buildings by certification systems such as the German Sustainable Building Assessment System (BNB) with the help of IFC-based building models. The automated provision of information from IFC models for the calculation of life cycle assessments (LCA), especially with information from the field of MEP, makes it possible to take into account aspects that have so far only been considered in an overview due to the high effort involved. In this way, optimization potentials can be better recognized and the lifecycle assessment can be used as an early and iterative planning tool.

Although the IFC4 data model offers the basic ability to integrate various environmental indicators for LCA and their units, sufficient conformity with DIN EN 15804 (NN 2020f) is not given. In the project, an Information Delivery Manual (IDM) with the required exchange processes and data transfer points, as well as a Model View Definition (MVD) with all the necessary classes and feature lists were created on the basis of the IFC conventions, in order to supply Life Cycle Assessment software automatically with the necessary information. Furthermore, the IFC supplements also include options for returning the LCA results in IFC format to the digital building models. To ensure performance and manageability of the building models despite the additional LCA information requirements, the project also pursued a multimodel container approach according to the linked data method explained in Sect. 10.2.4.

The influence of FM processes of buildings and properties (facility services—FS) on sustainability and the contribution to carbon emissions of the building is considered in some certification systems, such as the aforementioned BNB, by means of quality of process consideration, but this aspect is only taken into account to 10% and in a greatly simplified manner. Research results, among others, by Pelzeter et al. (2020) from the research project CarMa (Carbon Management for Facility Services) show that the contribution of FS to the carbon footprint (CFP) of a building requires a much more detailed consideration. In the *CarMa* project, a calculation method was developed to determine the CFP of FS according to the regulations of LCA based on ISO 14040. The method of carbon management also includes so-called Product Category Rules (PCR) required for calculation according to ISO 14025, which is documented in the German GEFMA guideline 162-1 (NN 2020c). In the project, a prototype LCA software tool was developed with which the CFP of all services required for building operations according to GEFMA 162-1 can be determined in a database-supported manner. The resource consumption required for FS is differentiated into the modules equipment, operating materials as well as the aspects of transport (e.g. of employees to their workplace) and overhead (e.g. cross-object support services) to be particularly considered for the provision of services. The tool has product catalogs for this purpose, which contain manufacturer information on the CFP of equipment and operating materials used, so-called Environmental Product Declarations (EPD). Furthermore, approxmation functions are offered if no EPD's are available from manufacturers, which is unfortunately still often the case, especially in the field of FS.

Since 2020, GEFMA and an association of FM service companies "The Enablers" have been working on a follow-up project at the HTW Berlin to further develop the resulting LCA software for facility services into a publicly accessible, open platform *carbonFM* (Krämer et al. 2021; NN 2021aa). In addition to project management of CFP projects, collaboration functions with role-based authorization management, the aforementioned product catalogs and evaluation functions for the identification of optimization approaches, carbonFM also offers ease of input effort to determine the CFP of FS by providing template projects, quality assurance functions and the development of parametric, intelligent service components, so-called smart service parts (SSP). SSP's, for example, allow the CFP overhead of a facility manager to be determined once using a few parameters and then reuse it frequently in the CFP project.

The development goal of *carbonFM* is initially to provide a tool for carbon management of FS in the sense of a self-assessment and internal benchmarking for FM service providers and consultants as well as RE managers. In perspective, the tool offers the potential to consider the CFP or carbon management of FS as part of FM tendering processes.

10.2.7 Smart Buildings and IoT

Smart buildings and the Internet of Things (IoT) are very extensive research areas. An exemplary project from Switzerland is presented here.

The state-funded research project ZHELIO (User Assistance Systems for Smart Commercial Buildings) is currently intensively dedicated to this topic, involving the ZHAW with its institutes: Institute of Facility Management (IFM), Institute of Embedded Systems (InES) and Center for Product and Process Development (ZPP) as research partners and Leicom AG as an industrial partner.

The project explores how data from buildings and sensors can be intelligently linked in specific Use Cases in order to enable building operators and users to save costs, reduce energy consumption, and enable digital, social interactions. The use cases are tailored to the end user so that every user can interact with the building infrastructure via so-called digital touchpoints. More than 100 of these use cases have now been identified and summarized by the research team into four use case clusters. Within these clusters, selected use cases are then implemented as a *proof of concept* in the office and school building of the IFM (RA building) at ZHAW in order to enable further research after the completion of the Innosuisse project.

The new platform *ZHELIO* emerging from the research work is intended to provide building operators with seamless and timely information in order to improve comfort, space and resource utilization, and productivity, and to help organizations achieve their demanding goals in a sustainable manner.

These goals are:

1. *Digital Twin and BIM*
   The research focuses on a user-assistance solution that enriches basic information from the construction planning phase with dynamic live data. In concrete terms, construction and operational data are brought together. The resulting model will give an insight into the dynamic state of the building and facilitate individual user guidance for all types of building users. Mobile devices and voice control will be essential for users to interact with various touchpoints.
2. *Digital Touchpoints*
   The use cases to be implemented as part of the research project will communicate and exchange data with building users via many digital touchpoints. The individually tailored information allows to experience a digital user journey within the interacting building. This results in many different digital touchpoints for operators and users of an infrastructure. Fig. 10.1 contains some specific touchpoints, but these can be extended at any time. The central element here is the simplification of processes. For example, in FM, new services and innovative solutions can be offered. Today, for example, the complaint of a building user means up to 10 more actions with corresponding interfaces. Often expensive experts of technical services for specialized applications must be involved.
   The research work considers the effects of a smart building on FM. Digital services route infrastructural services directly via a central interface. Much is automated and does not require the interaction of a specialist anymore. Combined with artificial intelligence and modern communication technology, a larger service scope can be offered with fewer resources.
3. *An App vs. 30 Apps*
   Infrastructures speak countless different "languages" and require a specific expert system for each application to control and evaluate. Most of these systems (e.g. BMS, energy monitoring, workplace analytics, CAFM) are also arranged as silos and often do not communicate with other systems.
   In a smart building, it is therefore essential to implement an overarching system as a platform for all individual expert systems. All data from their silos is transferred to a

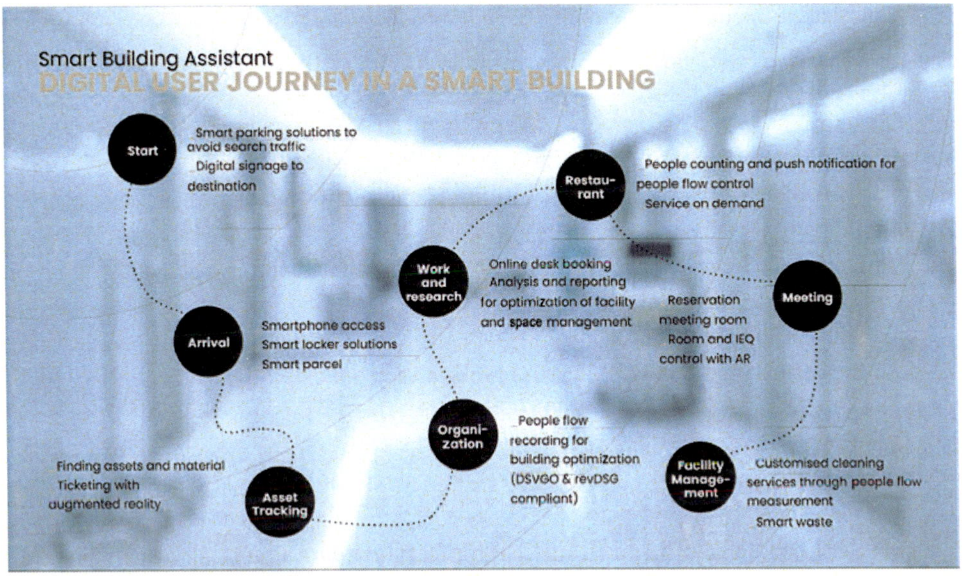

Fig. 10.1 Digital journey with different touchpoints through a smart building

data landscape via common interfaces and considered uniformly. This allows the FM, for example, to recognize data relationships, derive specific actions from them or enrich this pool with additional data in order to optimize costs, processes and personnel.

4. *Example Use Case: Indoor Environment Quality* (IEQ)
   Challenge:
   The comfort at the workplace is directly related to the productivity of the team. Aggregated monitoring and targeted data collection of thermal conditions, lighting, acoustics and ventilation are often difficult or impossible to achieve through traditional building automation.
   Solution approach:
   In a smart building app, the IEQ values of various, independent expert systems are summarized and displayed via a user interface, e.g. in an individually configurable dashboard. The quality of the indoor climate can be comprehensively assessed with all the values and representations available in the building, including derived and calculated values. By maintaining all these parameters in a standardized and rule-based manner, the overall IEQ is influenced and improved.
   Added value:
   The well-being, health and work performance of the team are significantly influenced by an improved IEQ. The control of the individual IEQ values by the users at the workplace or via their smartphones also allows for individual adjustments. Furthermore, data can be collected, evaluated and trends can be identified.
   Result:

The goal of the use case is to increase the productivity of employees through targeted IEQ optimization. In practice, productivity increases of 8-15% can be demonstrated (Schädlich et al. 2006; NN 2021ai).

Conclusion:

The individual digital journey through a building should actively support users in many of their activities, or at least in the background, in order to improve everyday life in many ways.

With a comprehensive assessment of the indoor climate and the occupancy of a room, it will not only be possible to react to changes in indoor climate, but also to act predictively in the event of changes in the room, such as the number of people, and to adjust the air volume flow so that CO_2 values vary only slightly.

10.3 Summary

The biggest benefit of BIM to FM is that the basic data of a building to be managed is available digitally after completion and does not have to be captured anew. This means that all relevant data for management is stored in a building data model and the current as-built status can be retrieved and used at any time.

The integration of CAFM, BIM and other software is a key to success for the digitalization of the lifecycle of real estate. The approach for this is provided by collaboration tools such as the developing model servers (CDE). Thus, it is only necessary to connect the various software systems to the BIM server and then all participants can work on a common data model. This model is centrally made available to all participants (on the Internet). Data security and encryption are then mandatory tasks. The CAFM system of the facility manager can use and edit all relevant data for the services to be provided. These data are stored in the BIM model versioned. Since building-related data is stored in the central BIM model, it is secondary which software is used. As long as the BIM model or the model server is supported, the software suitable for the task can be used.

In 2015, the German federal government developed a roadmap for the digitalization of planning and construction (2015c) which led to the recommendation of the use of BIM for public tenders from 2021 onwards. This will also increase the pressure on CAFM vendors to take over selected BIM data in their software. As the digitalization progresses in construction and real estate management, BIM will continue to play an important role. While some CAFM vendors see the BIM hype subsiding, it has now established itself as a standard with most major software providers (cf. Sect. 8.2) and is being further developed. The development towards a standard will greatly simplify the interface configuration.

The added value of BIM data becomes quickly apparent for new innovative buildings, for example when indoor navigation or energy simulations are carried out on the basis of BIM models, or when structured information about the building is available. It is assumed that BIM will continue to develop and increasingly establish itself in RE and FM.

The chapter first shows this perspective for BIM in RE management and FM with a critical introductory review. The number of product announcements, events and publications in the field of BIM far outweighs the number of implemented projects. Sect. 10.1 shows typical stumbling blocks, for example when considering the economic viability of BIM applications and the effort required for the continuous updating of as-built BIM data, but also refers to opportunities and positive developments for the use of BIM.

Sect. 10.2 shows, quite logically, what current and recent research has to offer in order to remove existing obstacles to the use of BIM in RE and FM and to make future potentials usable. Starting from an overview of the most important initiatives in the field of standardization, the section first discusses research activities in digital capturing of existing buildings. For example, approaches to (partially) automated processing of 3D point clouds, as they result from laser scans or photogrammetric surveys, are presented for the creation of BIM models of existing buildings (Scan2BIM).

The second research area presented relates to the management of BIM models in the operational phase. Innovative approaches for a common data environment (CDE) for RE and FM based on virtual integration according to linked data approaches are explained, as well as the development of open platforms for the selection and support of continuous, digital processing chains (tool chains) for BIM processes.

With research approaches in the field of visualization and virtual or augmented reality, another research area is presented using selected projects, where first research results are already available in practice in the field of maintenance management. Model-based approaches, as offered by the BIM method, help to bridge the gap between planning, construction and later operations. This opens up new possibilities for providing know-how from FM more easily and quickly for planning. The section shows research initiatives from the development of assistance systems to a framework for BIM-based knowledge management systems. It is not only a question of making knowledge from real estate and facility management usable, but also of efficiently and practically conveying it to practitioners, trainees and students. For this purpose, new opportunities for the use of 3D gaming environments (serious games) are presented, thus rounding off the field of knowledge management in RE and FM.

The very comprehensive research focus on sustainability, energy efficiency and CO_2 optimization is exemplarily treated with two research initiatives. Thus, approaches for the use of BIM models to create life cycle assessments more easily according to certification systems such as the German Sustainable Building Assessment System with the help of IFC-based building models are presented. Another research initiative concerns the development of an open platform with which the carbon footprint of facility services in real estate operations can be calculated, compared in benchmarks and finally optimized.

Sect. 10.2 concludes with the research and testing of future scenarios for smart buildings in order to raise the interaction between buildings and building users to a new level with novel IoT- and BIM-based services. Thus, an open platform for the integration of various app approaches is shown with the aim of increasing the productivity of building users by a significantly improved and measurable "indoor environment quality".

References

Bartels N (2020) Strukturmodell zum Datenaustausch im Facility Management. Baubetriebswesen und Bauverfahrenstechnik, Dissertation Technische Universität Dresden, Springer Vieweg, p 42 ff

Besenyöi Z, Kraemer M (2021) Towards the Establishment of a BIM-supported FM Knowledge Management System for Energy Efficient Building Operations. Proc. of the 38th International Conference of CIB W78, Luxembourg, 13–15 October, 194–203. http://itc.scix.net/paper/w78-2021-paper-020

Hallermann N, Debus P, Taraben J, Benz A, Morgenthal G, Rodehorst V, Völker C, Abbas T, Gebhardt T, Dauber S (2019) Unbemannte Fluggeräte zur Zustandsermittlung von Bauwerken – Fortsetzungsantrag. Abschlussbericht Forschungsinitiative Zukunft Bau, Band F 3157; Fraunhofer IRB Verlag

Krämer M, Besenyöi Z, Sauer P, Herrmann F (2018) Common Data Environment für BIM in der Betriebsphase – Ansatzpunkte zur Nutzung virtuell verteilter Datenhaltung. In: Bernhold T, May M, Mehlis J: Handbuch Facility Management, ecomed SICHERHEIT Verlag, Heidelberg, München, Landsberg, Frechen, Hamburg, pp 1–32

Krämer M, Besenyöi Z (2018) Towards Digitalization of Building Operations with BIM. In: IOP Conference Series: Materials Science and Engineering, IOP Publishing Ltd, Moskau, pp 1–11

Krämer M, May M, Salzmann P (2021) FM's Carbon Footprint – First Compute, then Improve. FMJ (USA), 31(November/December 2021)6, 58–61

Lambertz M, Theißen S, Höper J, Wimmer R, Mein-Becker A, Zibell, M (2019) Ökobilanzierung und BIM im Nachhaltigen Bauen. Endbericht. Bundesamt für Bauwesen und Raumordnung – (BBR), Bundesinstitut für Bau-, Stadt- und Raumforschung (BBSR), Forschungsprogramm Zukunft Bau, Bonn, 47 p

May M (2013) Serious Play – Computer Game Facilitates FM Learning. FMJ (USA), 23(September/October 2013)5, 23–27

May M (2017) BIM-based Augmented Reality for FM. FMJ (USA), 27(March/April 2017)2, 16–21

May M, Bernhold T (2009) FM-Assist: Tastendruck statt Berater? Immobilien Zeitung Nr. 37, 17.09.2009, 14

May M, Clauss M, Salzmann P (2017) A Glimpse into the Future of Facility and Maintenance Management: A Case Study of Augmented Reality. Corporate Real Estate Journal 6(2017)3, 227–244

Motawa I, Almarshad A (2013) A knowledge-based BIM system for building maintenance. Automation in Construction 29(2013) 173–182

NN (2009a) GEFMA Richtlinie 110: Einführung von Facility Management – Vorgehen bei der FM-Einführung in Unternehmen und öffentlichen Verwaltungen, Januar 2009, 4 p

NN (2015c) Stufenplan zur Einführung von BIM, Endbericht 31.12.2015, BMVI. https://www.bmvi.de/SharedDocs/DE/Anlage/DG/Digitales/bim-stufenplan-endbericht.pdf?__blob=publicationFile (retrieved: 11.12.2021)

NN (2019l) BIM-basiertes Betreiben. Bergische Universität Wuppertal. https://biminstitut.uni-wuppertal.de/de/forschung/abgeschlossene-forschungsprojekte/bim-basiertes-betreiben.html (retrieved: 13.12.2021)

NN (2020c) GEFMA-Richtlinie 162-1: Carbon Management von Facility Services. Januar 2020, 14 p

NN (2020e) DIN EN ISO 23386. Bauwerksinformationsmodellierung und andere digitale Prozesse im Bauwesen – Methodik zur Beschreibung, Erstellung und Pflege von Merkmalen in miteinander verbundenen Datenkatalogen. Deutsches Institut für Normung, 2020-11, 53 p

NN (2020f) DIN EN 15804. Nachhaltigkeit von Bauwerken – Umweltproduktdeklarationen – Grundregeln für die Produktkategorie Bauprodukte. Deutsches Institut für Normung, 2020-03, 81 p

NN (2021t) IBPDI – International Building Performance & Data Initiative. https://ibpdi.org/ (retrieved: 21.09.2021)

NN (2021x) BIM2TWIN – Optimal Construction Management & Production Control. https://cordis.europa.eu/project/id/958398/de (retrieved: 25.09.2021)

NN (2021y) BIM2digitalTWIN – Digitalisierung von Shopping-Centern – Von BIM zum Digital Twin. https://biminstitut.uni-wuppertal.de/de/forschung/abgeschlossene-forschungsprojekte/bim2digitaltwin.html (retrieved: 25.09.2021)

NN (2021z) BIMSWARM – SoftWare reference ARchitecture for openBIM. https://www.bim-swarm.de (retrieved: 25.09.2021)

NN (2021aa) carbonFM-Plattform. https://carbonfm.de/ (retrieved: 13.12.2021)

NN (2021ah) Forschungsprojekt BIMKIT. https://bimkit.eu/ (retrieved: 22.10.2021)

NN (2021ai) Der Einfluss des Raumklimas auf die Produktivität der Mitarbeiter. https://www.oxy-com.com/de/blog-nachrichten/der-einfluss-des-raumklimas-auf-die-produktivit%C3%A4t-der-mitarbeiter (retrieved: 22.10.2021)

NN (2021az) BIMeta. Plattform zur Verwaltung von Klassen und Merkmalen für den offenen BIM-Datenaustausch. https://www.bimeta.de/ (retrieved: 11.12.2021)

Pelzeter A, May M, Herrmann T, Ihle F, Salzmann P (2020) Decarbonisation of Facility Services Supported by IT. Corporate Real Estate Journal 9(2020)4, 361–374

Salzmann P, May M (2016) Mehr Durchblick mit Augmented Reality. Jahrbuch Facility Management 2016. Der F.A.Z.-Fachverlag, Friedberg, Februar 2016, 118–123

Schädlich S, Röttger I, Lüttgens S (2006) Menschliche Behaglichkeit in Innenräumen und deren Einfluss auf die Produktivität am Arbeitsplatz. Studie der Fritz-Steimle-Stiftung, August 2006, 63 p

Appendix 1: Checklist for Implementing BIM in FM

11

Thomas Bender and Matthias Mosig

This checklist contains the essential tasks for the introduction of BIM in FM. The contents described below are to be understood as a guideline and concrete assistance and are intended to make a significant contribution to a successful BIM-in-FM implementation.

Where possible, the tasks have been arranged chronologically and are based on the professional and content-related explanations in this book.

No.	Task	Time	Involved
1	Ensure familiarization with BIM ISO 19650 standards and developing an overarching BIM4FM strategy. Define what is the stakeholder expectation of BIM, BIM4FM? Describe which goals should be pursued (cost, time, quality)? Identify which processes and activities can be designed more efficiently and effectively through the BIM methodology (processes, roles, tools, data, etc.)? Clarify which added value is to be created to help manage expectations? Confirm how (processes, roles, tools etc.) should the described goals be achieved?	Preparation & Basics	Owner, Operator, Facility Manager

T. Bender (✉)
pit – cup GmbH, Heidelberg, Germany
e-mail: thomas.bender@pit.de

M. Mosig
TÜV SÜD Advimo GmbH, Munich, Germanye-mail: matthias.mosig@tuvsud.com

M. May et al. (eds.), *BIM in Real Estate Operations*,
https://doi.org/10.1007/978-3-658-40830-5_11

No.	Task	Time	Involved
2	Building up BIM4FM expertise in the organization, e.g. as a client BIM information manager with a focus on FM. The role should be a permanent part of the BIM project team and the later FM organization. Develop required process know-how (from the perspective of planning, construction and operation). Ensure IT know-how (in particular CAFM, CDE and authoring tools). Ensure there is adequate knowledge of the required information in the relevant processes.	Operation & Use	Client BIM Information Manager (with FM focus)
3	Description of the specific requirements for a BIM project (new construction, renovation, etc.) and BIM-based data management in the operational phase from an FM perspective. The result is the project information requirements (PIR), which are to be integrated into the BIM project via EIR and BEP (cf. Sect. 7.2 "Procedure in a BIM project"). Key contents of the PIR: – Description of the BIM application cases required for the respective processes (e.g. maintenance) with data and information (target definition, what is to be handed over) – Graphic data (model content → LOG) – Semantic information on the objects (property sets → LOI) – Documents on the objects – Transfer formats – Transfer times – Description of the IT infrastructure into which the data are to be integrated after the construction phase – Description of the requirements for the processes for implementing the BIM application cases for the planning, construction and operational phases	Preparation & Basics	Owner, Operator, Facility Manager, Client BIM Information Manager (with FM focus)
4	Establishing the BIM information manager in the BIM project, e.g. with the following activities: – Participation in BIM meetings – Requirement definition – Quality assurance – Data transfer Ideally, this role and staffing is also responsible for BIM data management in the operational phase.	Across phases	Client BIM Information Manager (with FM focus)
5	Participate in the creation of EIR and BEP: – Integration of FM requirements (PIR) – Comparison with specific project requirements and -conditions	Preparation & Basics, Planning	Client BIM Information Manager (with FM focus)

No.	Task	Time	Involved
6	Contribute to the selection, implementation and operation of the relevant BIM tools in the project (CDE, BIM database, etc.).	Preparation & Basics	Client BIM Information Manager (with FM focus)
7	Contribute to the implementation of a quality assurance process in the BIM project and to changes in the operational phase. Training of all relevant stakeholders with regard to the requirements of the BIM use cases. The quality check is carried out in coordination with the BIM manager and the BIM overall coordination. Check the BIM information model (geometry and semantics) for compliance with the agreed delivery services. Check at the defined milestones in the project (from the design planning to the handover of the as-built model).	Aross phases	Client BIM Information Manager (with FM focus)
8	Providing a suitable BIM information model for the tendering of FM services (cleaning, maintenance, etc.).	Execution	Client BIM Information Manager (with FM focus)
9	Participate in the commissioning phase. Check the final delivery (as-built models and all associated information/data inside but also outside of models). Handover/integration of the as-built model(s) and information/data into the operational phase (integration and use in CAFM and other FM management systems, e.g. ERP).	Project completion	Client BIM Information Manager (with FM focus), Facility Manager
10	Continuation and maintenance of the BIM information models. Ensuring the communication of changes such as repairs, redevelopment, conversion, and renovations. Updating of contents at a central point (graphically, alphanumerically, digital documents). Provision of inventory models in the event of conversion measures or direct access to the central storage (e.g. CDE in the operating phase). Merging of changes at a central point (unless direct access and editing via a CDE).	Operation & Use	Client BIM Information Manager (with FM focus), Facility Manager User

Appendix 2: Overview of Standardization Initiatives

12

Matthias Mosig and Marko Opić

The following initiatives pursue the standardization of data or exchange formats in the BIM2FM environment with different approaches and degrees of intensity. This extract does not claim to be complete and is intended to serve as a guide. In addition, there are numerous other initiatives, such as from software vendors' associations with very specific product standards.

No.	Designation	Short description	Web link
1	GEFMA	Comprehensive set of guidelines, inter alia, for the implementation and optimization of CAFM/IWMS systems (guideline series 400 ff) and for data and documents in the life cycle of FM (guideline series 922-1 ff, 924 ff, 926). Publisher of various white papers on the topic of digitalization in FM.	https://www.gefma.de/

M. Mosig (✉)
TÜV SÜD Advimo GmbH, München, Germany
e-mail: matthias.mosig@tuvsud.com

M. Opić
Alpha IC GmbH, Nürnberg, Germany
e-mail: m.opic@alpha-ic.com

M. May et al. (eds.), *BIM in Real Estate Operations*,
https://doi.org/10.1007/978-3-658-40830-5_12

No.	Designation	Short description	Web link
2	RealFM	Publisher of a BIM guideline and provider of seminars for training on the BIM guideline as a basis for the implementation and application of BIM in FM. Development of an MEP parameter list from the perspective of calculation relevance, maintenance relevance, testing obligation.	https://www.realfm.de/
3	VDI	Guideline series VDI 2552 sheet 1-11, VDI 3805 on BIM basics, terms, model-based quantity takeoff for cost planning, scheduling, tendering and billing, data exchange, data management, processes, qualifications—basic knowledge, classification systems, employer information requirements (EIR) and BIM execution plans (BEP), information exchange requirements—formwork and scaffolding technology (in-situ concrete construction).	https://www.vdi.eu/
4	ISO/DIN	ISO 19650 series of standards for the definition, exchange, organization and processing of information.	https://www.iso.org
5	CEN	CEN/TC 442 series of standards for BIM.	https://www.cen.eu/Pages/default.aspx
6	BSI—PAS	PAS 1192-2:2013 series of standards replaced by BS EN ISO 19650-1 Organization of information about construction works—Information management using building information modeling—Part 1: Concepts and principles and BS EN ISO 19650-2 Organization of information about construction works—Information management using building information modeling—Part 2: Facilities delivery phase.	https://www.bsigroup.com/de-DE/Ueber-BSI-Group/
7	ÖNORM	ÖNORM A 7010-6 Object management—Part 6: Requirements for data from BIM models over the life cycle.	https://www.austrian-standards.at/
8	BIMETA	Open digital platform of the construction and MEP industry for all relevant BIM objects. Provision of BIM templates with reference to relevant regulations, guidelines and standards as well as to the buildingSMART Data Dictionary (bSDD) for all essential product data.	https://www.bimeta.de/

No.	Designation	Short description	Web link
9	DIN BIM Cloud	German cloud-based BIM content database with standardized component properties and networking with the international and national construction standards world. The content is machine-readable and technically compatible with the recognized rules of technology. There is also a link, for example, with STLB-Bau, DIN 276 and IFC.	https://www.din-bim-cloud.de/
10	IBPDI	International Building Performance & Data Initiative to define a Common Data Model as an open standard for all real estate-related business processes, taking into account national and international standards for data exchange.	https://ibpdi.org/about/
11	CAFM-Connect	Open BIM interface of the German CAFM RING for the exchange of real estate data in planning, construction, operation with an open data standard, based on IFC. Provided by BIM profiles as data exchange standards for BIM use cases in the operation of buildings. The CAFM-Connect editor allows capture of buildings, their components and documents based on BIM profiles.	https://www.cafm-connect.org/
12	BIMSWARM	IT platform for the digitalization of planning, construction and operation with a focus on BIMSWARM marketplace, BIMSWARM certification, compatibility of construction IT products, market intelligence and user assessments as well as neutrality of the platform operator.	https://www.bimswarm.de/
13	Industry Foundation Classes (IFC)	Open interface as a data model for the exchange of model-based information between different software applications throughout the entire lifecycle of real estate. Available as a separate standard ISO 16739 since IFC4.	https://www.buildingsmart.de/bim-knowhow/standards-standardisierung
14	NBIMS-US and NCS	Open national BIM standards in the USA related to existing standards for facility and construction planning as well as operational concepts.	https://www.nibs.org/resources/standards
15	COBie	Part of the national BIM standards NBIMS-US and the British BS 1192-4, used to exchange alphanumeric building data with a focus on FM.	https://www.bsigroup.com/de-DE/

No.	Designation	Short description	Web link
16	gif IDA Model	Directive for the exchange of real estate data with all the necessary data fields for successful cooperation of market participants in real estate management on the basis of a process and hierarchical entity relationship model as well as XML schemas of the structure of the XML documents.	https://www.zgif.org/de/
17	gbXML	Green Building eXtended Markup Language, enables the exchange of data between 3D CAD / BIM systems and technical calculation programs and analysis tools.	https://www.gbxml.org/
18	OmniClass	Comprehensive classification system for the construction industry for classifying the entire built environment over the project lifecycle.	https://www.csiresources.org/standards/omniclass
19	Real Estate Core	Swedish open-source building ontology that prepares buildings for interaction with the smart city by combining existing standards.	https://www.realestatecore.io/
20	Brickschema	Open-source building ontology for standardizing semantic descriptions of physical, logical, and virtual assets in buildings and the relationships between them.	https://brickschema.org/
21	Project Haystack	Open-source building ontology with a technical focus, driven by MEP manufacturers to optimize the processing of IoT data.	https://project-haystack.org/

List of Figures

Figure	Source
1.1	United Nations Department of Global Communications. The content of this publication has not been approved by the United Nations and does not reflect the views of the United Nations or its officials or Member States.
1.2, 1.4	Baldegger et al. 2021
1.5	Swiss Federal Railways SBB
1.8	Patacas et al. 2020
1.9	LIBAL Switzerland GmbH
2.5	TÜV Süd Advimo
2.12–2.13	Axonize
2.15	Christian Müller, DFKI
2.16–2.18	RECOTECH GmbH
2.19–2.20	Messer Construction and Xavier University
2.21	Trzechiak 2017
2.22	Gruschke and Werner 2013
2.23	Verena Rock, TH Aschaffenburg
3.4	based on Sacks et al. 2018
3.5	PB P. Berchtold Ing. HTL/HLK Ingenieurbüro für Energie & Haustechnik
3.6	Planon
3.8	based on Borrmann et al. 2018, p. 13
3.9	based on BIM Dimensionen, Höflich & Maier Consult GmbH
3.13	CAFM Ring e.V.
3.14–3.16	based on Building and Construction Agency, Singapore
4.1	Planon and Cadac
4.2	Planon
4.3	Archibus Solution Center Germany

M. May et al. (eds.), *BIM in Real Estate Operations*,
https://doi.org/10.1007/978-3-658-40830-5

Figure	Source
4.5	Bundesministerium für Verkehr und digitale Infrastruktur, BIM4Infra2020, Teil 10
4.8	Krämer, based on NN 2019i
4.9, 4.12	Planon
4.14–4.15	using IFC model from Institute for Automation and Applied Informatics / Karlsruhe Institute of Technology
5.1	The figure was created using resources from flaticon.com
7.4	Federal Ministry of Transport and Digital Infrastructure, BIM4Infra2020, Part 6
8.3–8.5	GEFMA e.V.
8.6	SAP SE
8.8	IBPDI
8.10	BuildingMinds
9.1 (left)	Axel Springer SE
9.1 (right)	Elisabeth May
9.3	Ed. Züblin AG
9.5	Archimation on behalf of MfN
9.6	based on AIA MfN
9.9–9.10	ProSiebenSat.1 planning documents
9.14	IBP, Singapore
9.18–9.24	BPS International GmbH
9.25–9.31	Archibus Solution Center Germany
9.32	Tempelhof Projekt GmbH
9.35	Ingo Rasp, ingorasp.com
9.36	pom+Consulting AG
9.37	LIBAL Schweiz GmbH
9.38–9.40	Hochbauamt Graubünden
10.1	Leicom AG, 2021

We would like to thank all the people, companies and organizations listed for their permission to reprint the figures in this book.

All other images are from the authors of the respective chapters or sections.